Spon's Estimating Costs Guide to Plumbing and Heating

Other books by Bryan Spain

also available from Taylor & Francis

Spon's Estimating Costs Guide to Electrical Works (2006 edition)
Pb: 0-415-38614-4

Spon's Estimating Costs Guide to Minor Works, Alterations and Repairs to Fire, Flood, Gale and Theft Damage (2006 edition)
Pb: 0-415-38213-0

Spon's Estimating Costs Guide to Minor Landscaping, Gardening and External Works (2005 edition)
Pb: 0-415-34410-7

Spon's Estimating Costs Guide to Roofing (2005 edition)
Pb: 0-415-34412-3

Spon's Estimating Costs Guide to Finishings: painting and decorating, plastering and tiling (2005 edition)
Pb: 0-415-34411-5

Spon's House Improvement Price Book: house extensions, storm damage work, alterations, loft conversions and insulation (2005 edition)
Pb: 0-415-37043-4

Spon's First Stage Estimating Price Book (2006 edition)
Pb: 0-415-38619-5

Spon's Construction Resource Handbook (1998 edition)
Hb: 0-419-23680-5

Information and ordering details

For price availability and ordering visit our website
www.tandf.co.uk/builtenvironment

Alternatively our books are available from all good bookshops.

Spon's Estimating Costs Guide to Plumbing and Heating

Unit rates and project costs

Third edition

Bryan Spain

LONDON AND NEW YORK

First published 1999 by Spon Press
Second edition 2004
Third edition published 2006 by Taylor & Francis
2 Park Square, Milton Park, Abingdon, Oxon OX14 4RN

Simultaneously published in the USA and Canada
by Taylor & Francis
270 Madison Avenue, New York, NY 10016, USA

Reprinted 2006

Routledge is an Imprint of the Taylor & Francis Group, an informa business

Publisher's note
This book has been prepared from camera-ready copy supplied by the author.

Printed and bound in Great Britain by TJ International Ltd, Padstow, Cornwall.

British Library Cataloguing in Publication Data
A catalogue record for this book is available from the British Library

ISBN 10: 0-415-38618-7
ISBN 13: 9-78-0-415-38618-0

Contents

Bathrooms

Preface

A small to medium-sized plumbing business competing in today's commercial world needs all the help it can get. The pressure of obtaining work, carrying it out efficiently and ensuring that full payment is made on time is the responsibility of the firm's principal. Any source of information that can help in saving time must be welcome and this book aims to fulfil this need.

Plumbers and heating engineers have had access to price books for many years but they were always obliged to prepare their own quantities and extract the relevant priced items from the books in order to prepare an estimate. This procedure is time-consuming and it is particularly galling when the contractor suspects that the enquiry is not genuine but merely a ruse to check a previously accepted estimate.

This book provides detailed costs of projects in addition to setting out analytical unit rates. It is accepted that it is unlikely that the projects selected will be identical in size to those received in actual enquiries but the range of projects covered should enable a contractor to provide a quick and accurate response to an enquiry by making slight adjustments to the examples provided.

A detailed quotation could then be prepared if the client showed interest in the estimate. This should make an appreciable saving in the one commodity that all contractors are short of...time!

I have received a great deal of support in the research necessary for this type of book and I am grateful to those individuals and firms who have provided the cost data and other information. In particular, I am indebted to Mark Loughrey of Loughrey & Co., Chartered Accountants of Hoylake (tel: 0151-632 3298 or www.yesl.uk.com) who are specialists in advising small businesses. Their research for the information in the business section is based on tax legislation in force in July 2005.

Although every care has been taken in the preparation of the book, neither the publishers nor I can accept any responsibility for the use of the information provided by any firm or individual. Finally, I would welcome any constructive criticism of the book's contents and suggestions that could be incorporated into future editions.

Bryan Spain
www.costofdiy.com
July 2005

Introduction

This book is intended to help small to medium-sized plumbing businesses to carry out their estimating work more quickly and more accurately. It is hoped that the resulting saving time will enable contractors to run their businesses more efficiently.

The contents of this book cover unit rates, project costs and general advice on business matters. The unit rates section presents analytical rates for all types of work encountered in domestic and light industrial plumbing contracts. Some rates for alterations and repairs are also included.

The project costs section contains the total costs for carrying out the following types of work and each of these categories is sub-divided into different sized projects and types of materials:

- rainwater goods
- bathrooms
- external waste systems
- central heating systems
- hot and cold water systems.

The business section covers advice on starting and running a business, together with information on taxation and VAT matters.

Rainwater goods

This section lists the quantities and costs of PVC-U, cast iron and aluminium rainwater gutters and pipes for 18 different types and sizes of houses. These range from a one-storey terraced house, overall size 7 × 6m with gable ends, to a three-storey house, overall size 10 × 9m with hipped ends. Each house within the range is costed for PVC-U, cast iron and aluminium pipes and gutters to produce a total coverage of rainwater goods for 54 types of houses.

For example, on pages 91 and 92, the total cost of taking down the existing rainwater pipes and gutters and replacing them in PVC-U for a semi-detached two-storey house, overall size 8 × 7m with hipped ends is £562.44. This is broken down into £273.42 for labour, £215.66 for materials and £73.36 for overheads and profit. Due to rounding off, the vertical and horizontal totals may not always coincide exactly.

In the above example, the time allowed to do the work is shown as 17.64 man hours, i.e. 17 hours 38 minutes, but the actual period over which the work is carried out may be longer due to other factors. The totals of the projects are taken to a summary on pages 119 to 121.

Bathrooms

This section covers bathroom layouts with varying combinations of the following sanitary fittings:

- bath
- bath with hand shower set
- WC
- lavatory basin
- shower cubicle
- bidet.

A total of 16 bathrooms are included and the totals are summarised on pages 159 to 160.

External waste systems

Soil pipes and waste pipes from wash basins and baths are included covering four types of pipe for one-, two- and three-storey houses producing a combination of twelve different cases. These are summarised on page 171.

Central heating systems

Gas-fired central heating systems for three sizes of one-, two- and three-bedroom houses are included and summarised costs appear on page 191.

Hot and cold water systems

The project costs for the installation of hot and cold water pipes including storage tanks and cylinders, are covered for one-, two- and three-storey houses and the labour hours and costs are included on page 198.

Materials

One of the problems facing small plumbing and heating contractors is their inability to obtain large discounts on materials because the nature of the work that they undertake precludes them from buying in large quantities.

The discounts allowed in this book are generally 10-15%. The allowance for waste is 5% except for large items, such as boilers and radiators, where 1% has been included.

Labour

The labour rate has been set at £15.50 per hour for all work in this book and is based on current rates paid to Advanced Plumbers.

Overheads and profit

This has been set at 15% for all grades of work and is deemed to cover head office and site overheads, including:

- heating
- lighting
- rent
- rates
- telephones
- secretarial services
- insurances
- finance charges
- transport
- small tools
- ladders
- scaffolding etc.

Part One

UNIT RATES

H71	Lead sheet coverings
N13	Sanitary appliances/fittings
P31	Holes, chases, covers and supports for services
R10	Rainwater pipework/gutters
R11	Drainage above ground
R12	Drainage below ground
S10	Hot and cold water
T10	Gas/oil-fired boilers
T30	Radiators

Alterations and repairs

	Unit	Labour hours	Hours cost £	Materials £	O & P £	Total £
H71 LEAD SHEET COVERINGS						
Flat roofing, less than 10 degrees to the horizontal						
code 4	m2	2.90	44.95	19.61	9.68	74.24
code 5	m2	3.10	48.05	26.14	11.13	85.32
code 6	m2	3.30	51.15	28.86	12.00	92.01
code 7	m2	3.50	54.25	34.21	13.27	101.73
code 8	m2	3.80	58.90	38.56	14.62	112.08
Dormers, less than 10 degrees to the horizontal						
code 4	m2	3.65	56.58	19.61	11.43	87.61
code 5	m2	3.90	60.45	26.14	12.99	99.58
code 6	m2	4.15	64.33	28.86	13.98	107.16
code 7	m2	4.35	67.43	34.21	15.25	116.88
code 8	m2	4.65	72.08	38.56	16.60	127.23
Sloping roofing, 10 to 50 degrees to the horizontal						
code 4	m2	3.10	48.05	19.61	10.15	77.81
code 5	m2	3.30	51.15	26.14	11.59	88.88
code 6	m2	3.50	54.25	28.86	12.47	95.58
code 7	m2	3.70	57.35	34.21	13.73	105.29
code 8	m2	3.90	60.45	38.56	14.85	113.86
Dormers, 10 to 50 degrees to the horizontal						
code 4	m2	3.65	56.58	19.61	11.43	87.61
code 5	m2	3.90	60.45	26.14	12.99	99.58
code 6	m2	4.15	64.33	28.86	13.98	107.16
code 7	m2	4.35	67.43	34.21	15.25	116.88
code 8	m2	4.65	72.08	38.56	16.60	127.23
Vertical or sloping roofing, 10 to 50 degrees to the horizontal						
code 4	m2	3.30	51.15	19.61	10.61	81.37
code 5	m2	3.50	54.25	26.14	12.06	92.45

	Unit	Labour hours	Hours cost £	Materials £	O & P £	Total £
Vertical or sloping roofing, 10 to 50 degrees to the horizontal (cont'd)						
code 6	m2	3.70	57.35	28.86	12.93	99.14
code 7	m2	3.90	60.45	34.21	14.20	108.86
code 8	m2	4.10	63.55	38.56	15.32	117.43
Dormers, over 50 degrees to the horizontal						
code 4	m2	3.65	56.58	19.61	11.43	87.61
code 5	m2	3.90	60.45	26.14	12.99	99.58
code 6	m2	4.15	64.33	28.86	13.98	107.16
code 7	m2	4.35	67.43	34.21	15.25	116.88
code 8	m2	4.65	72.08	38.56	16.60	127.23
Flashings, horizontal, girth 150mm						
code 4	m	0.30	4.65	2.94	1.14	8.73
code 5	m	0.35	5.43	3.92	1.40	10.75
Flashings, horizontal, girth 240mm						
code 4	m	0.40	6.20	4.71	1.64	12.55
code 5	m	0.45	6.98	6.27	1.99	15.23
Flashings, horizontal, girth 300mm						
code 4	m	0.50	7.75	5.88	2.04	15.67
code 5	m	0.60	9.30	7.84	2.57	19.71
Flashings, sloping, girth 150mm						
code 4	m	0.40	6.20	2.94	1.37	10.51
code 5	m	0.45	6.98	3.92	1.63	12.53
Flashings, sloping, girth 240mm						
code 4	m	0.50	7.75	4.71	1.87	14.33
code 5	m	0.55	8.53	6.27	2.22	17.01
Flashings, sloping, girth 300mm						
code 4	m	0.60	9.30	5.88	2.28	17.46
code 5	m	0.65	10.08	7.84	2.69	20.60

	Unit	Labour hours	Hours cost £	Materials £	O & P £	Total £
Flashings, stepped, girth 150mm						
code 4	m	0.50	7.75	2.94	1.60	12.29
code 5	m	0.55	8.53	3.92	1.87	14.31
Flashings, stepped, girth 240mm						
code 4	m	0.60	9.30	4.71	2.10	16.11
code 5	m	0.65	10.08	6.27	2.45	18.80
Flashings, stepped, girth 300mm						
code 4	m	0.70	10.85	5.88	2.51	19.24
code 5	m	0.75	11.63	7.84	2.92	22.38
Aprons, horizontal, girth 240mm						
code 4	m	0.50	7.75	4.71	1.87	14.33
code 5	m	0.55	8.53	6.27	2.22	17.01
Aprons, horizontal, girth 300mm						
code 4	m	0.60	9.30	5.88	2.28	17.46
code 5	m	0.65	10.08	7.84	2.69	20.60
Aprons, horizontal, girth 450mm						
code 4	m	0.70	10.85	8.82	2.95	22.62
code 5	m	0.75	11.63	10.47	3.31	25.41
Aprons, sloping, girth 240mm						
code 4	m	0.60	9.30	4.19	2.02	15.51
code 5	m	0.65	10.08	5.58	2.35	18.00
Aprons, sloping, girth 300mm						
code 4	m	0.70	10.85	5.24	2.41	18.50
code 5	m	0.75	11.63	7.00	2.79	21.42
Aprons, sloping, girth 450mm						
code 4	m	0.80	12.40	7.86	3.04	23.30
code 5	m	0.85	13.18	10.47	3.55	27.19
Sills, horizontal, girth 240mm						
code 4	m	0.50	7.75	4.19	1.79	13.73
code 5	m	0.55	8.53	5.58	2.12	16.22

	Unit	Labour hours	Hours cost £	Materials £	O & P £	Total £
Sills, horizontal, girth 300mm						
code 4	m	0.60	9.30	5.24	2.18	16.72
code 5	m	0.65	10.08	7.00	2.56	19.64
Sills, horizontal, girth 450mm						
code 4	m	0.70	10.85	7.86	2.81	21.52
code 5	m	0.75	11.63	10.47	3.31	25.41
Cappings, horizontal, girth 240mm						
code 4	m	0.50	7.75	4.19	1.79	13.73
code 5	m	0.55	8.53	5.58	2.12	16.22
Cappings, horizontal, girth 300mm						
code 4	m	0.70	10.85	5.24	2.41	18.50
code 5	m	0.80	12.40	7.00	2.91	22.31
Cappings, horizontal, girth 450mm						
code 4	m	1.00	15.50	7.86	3.50	26.86
code 5	m	1.10	17.05	11.76	4.32	33.13
Hips, sloping, girth 240mm						
code 4	m	0.55	8.53	4.19	1.91	14.62
code 5	m	0.65	10.08	5.58	2.35	18.00
Hips, sloping, girth 300mm						
code 4	m	0.80	12.40	5.24	2.65	20.29
code 5	m	0.90	13.95	7.00	3.14	24.09
Hips, sloping, girth 450mm						
code 4	m	1.10	17.05	7.86	3.74	28.65
code 5	m	1.20	18.60	11.76	4.55	34.91
Kerbs, horizontal, girth 240mm						
code 4	m	0.60	9.30	4.19	2.02	15.51
code 5	m	0.70	10.85	5.58	2.46	18.89
Kerbs, horizontal, girth 300mm						
code 4	m	0.70	10.85	5.24	2.41	18.50
code 5	m	0.80	12.40	7.00	2.91	22.31

	Unit	Labour hours	Hours cost £	Materials £	O & P £	Total £
Kerbs, horizontal, girth 450mm						
code 4	m	1.00	15.50	7.86	3.50	26.86
code 5	m	1.10	17.05	11.76	4.32	33.13
Ridges, horizontal, girth 240mm						
code 4	m	0.60	9.30	4.19	2.02	15.51
code 5	m	0.70	10.85	5.58	2.46	18.89
Ridges, horizontal, girth 300mm						
code 4	m	0.70	10.85	5.24	2.41	18.50
code 5	m	0.80	12.40	7.00	2.91	22.31
Ridges, horizontal, girth 450mm						
code 4	m	1.00	15.50	7.86	3.50	26.86
code 5	m	1.10	17.05	11.76	4.32	33.13
Valleys, horizontal, girth 450mm						
code 4	m	1.00	15.50	7.86	3.50	26.86
code 5	m	1.10	17.05	11.76	4.32	33.13
Valleys, horizontal, girth 600mm						
code 4	m	1.20	18.60	11.77	4.56	34.93
code 5	m	1.30	20.15	15.85	5.40	41.40
Gutters, sloping, girth 450mm						
code 4	m	1.10	17.05	7.86	3.74	28.65
code 5	m	1.30	20.15	11.76	4.79	36.70
code 6	m	1.50	23.25	12.99	5.44	41.68
code 7	m	1.60	24.80	15.39	6.03	46.22
Gutters, sloping, girth 600mm						
code 4	m	1.30	20.15	11.77	4.79	36.71
code 5	m	1.50	23.25	15.85	5.87	44.97
code 6	m	1.70	26.35	17.32	6.55	50.22
code 7	m	1.80	27.90	20.53	7.26	55.69
Edges, welted						
code 4	m	0.30	4.65	0.00	0.70	5.35
code 5	m	0.30	4.65	0.00	0.70	5.35
code 6	m	0.50	7.75	0.00	1.16	8.91

	Unit	Labour hours	Hours cost £	Materials £	O & P £	Total £
Edges, beaded						
code 4	m	0.30	4.65	0.00	0.70	5.35
code 5	m	0.30	4.65	0.00	0.70	5.35
code 6	m	0.50	7.75	0.00	1.16	8.91
Dressing over slating and tiling						
code 4	m	0.20	3.10	0.00	0.47	3.57
code 5	m	0.25	3.88	0.00	0.58	4.46
code 6	m	0.30	4.65	0.00	0.70	5.35
code 7	m	0.35	5.43	0.00	0.81	6.24
Soakers, size 150 × 150mm						
code 3	nr	0.20	3.10	1.32	0.66	5.08
code 4	nr	0.25	3.88	1.49	0.80	6.17
Soakers, size 200 × 300mm						
code 3	nr	0.25	3.88	2.22	0.91	7.01
code 4	nr	0.30	4.65	2.57	1.08	8.30
Soakers, size 300 × 150mm						
code 3	nr	0.30	4.65	2.66	1.10	8.41
code 4	nr	0.35	5.43	3.07	1.27	9.77
Slates, size 400 × 400mm with 200mm high collar 100mm diameter						
code 4	nr	1.40	21.70	10.66	4.85	37.21
code 5	nr	1.50	23.25	11.48	5.21	39.94
Slates, size 400 × 400mm with 200mm high collar 150mm diameter						
code 4	nr	1.60	24.80	11.18	5.40	41.38
code 5	nr	1.70	26.35	11.57	5.69	43.61
Dots, cast lead						
code 3	nr	0.70	10.85	1.78	1.89	14.52
code 4	nr	0.75	11.63	1.92	2.03	15.58
code 5	nr	0.80	12.40	2.06	2.17	16.63

	Unit	Labour hours	Hours cost £	Materials £	O & P £	Total £
N13 SANITARY APPLIANCES/ FITTINGS						
Baths						
Acrylic reinforced bath size 1500 × 700mm complete with 2nr chromium plated grips, 40mm waste fitting, overflow with chain and plastic plug						
white	nr	2.75	42.63	135.77	26.76	205.15
Acrylic reinforced bath size 1700 × 700mm complete with 2nr chromium plated grips, 40mm waste fitting, overflow with chain and plastic plug						
white	nr	2.90	44.95	153.94	29.83	228.72
coloured	nr	2.90	44.95	167.65	31.89	244.49
Acrylic reinforced bath size 1700 × 800mm complete with 2nr chromium plated grips, 40mm waste fitting, overflow with chain and plastic plug						
white	nr	3.00	46.50	186.65	34.97	268.12
coloured	nr	3.00	46.50	204.52	37.65	288.67
Acrylic reinforced bath size 1700 × 800mm complete with 2nr chromium plated grips, 40mm waste fitting, overflow with chain and plastic plug						
white	nr	3.10	48.05	241.40	43.42	332.87
coloured	nr	3.10	48.05	252.19	45.04	345.28

	Unit	Labour hours	Hours cost £	Materials £	O & P £	Total £
Porcelain enamel standard gauge bath size 1700 × 700mm complete with 2nr chromium plated grips, 40mm waste fitting, overflow with chain and plastic plug						
white	nr	3.00	46.50	161.25	31.16	238.91
coloured	nr	3.00	46.50	175.79	33.34	255.63
Porcelain enamel heavy gauge bath size 1700 × 700mm complete with 2nr chromium plated grips, 40mm waste fitting, overflow with chain and plastic plug						
white	nr	3.10	48.05	168.59	32.50	249.14
coloured	nr	3.10	48.05	212.21	39.04	299.30
Porcelain enamel heavy gauge (shallow) bath size 1700 × 700mm complete with 2nr chromium plated grips, 40mm waste fitting, overflow with chain and plastic plug						
white	nr	3.10	48.05	248.67	44.51	341.23
coloured	nr	3.10	48.05	289.43	50.62	388.10
Bath accessories						
Bath panels, enamelled hardboard, fixing with chromium-plated dome headed screws, cutting to length						
end panel	nr	0.30	4.65	10..75	0.70	5.35
side panel	nr	0.50	7.75	12.96	3.11	23.82
Bath panels, moulded acrylic, fixing in position for trimming as required						
end panel	nr	0.30	4.65	13.67	2.75	21.07
side panel	nr	0.50	7.75	16.33	3.61	27.69

	Unit	Labour hours	Hours cost £	Materials £	O & P £	Total £
Angle strip polished aluminium, fixing with chromium-plated dome headed screws, cutting to length						
25 × 25 × 560mm long	nr	0.35	5.43	4.95	1.56	11.93
Basins and pedestals						
Wash basin, vitreous china size 510 × 410mm complete with pedestal, chromium-plated waste, overflow with chain and plastic plug						
white	nr	2.20	34.10	89.88	18.60	142.58
coloured	nr	2.20	34.10	97.82	19.79	151.71
Wash basin, vitreous china size 560 × 455mm complete with pedestal, chromium-plated waste, overflow with chain and plastic plug						
white	nr	2.25	34.88	82.45	17.60	134.92
coloured	nr	2.25	34.88	99.78	20.20	154.85
Wash basin, vitreous china size 610 × 470mm complete with pedestal, chromium-plated waste, overflow with chain and plastic plug						
white	nr	2.30	35.65	93.31	19.34	148.30
coloured	nr	2.30	35.65	100.87	20.48	157.00
Wash basin, vitreous china size 510 × 410mm complete with brackets, chromium-plated waste, overflow with chain and plastic plug						
white	nr	2.00	31.00	69.22	15.03	115.25
coloured	nr	2.00	31.00	79.48	16.57	127.05

	Unit	Labour hours	Hours cost £	Materials £	O & P £	Total £
Wash basin, vitreous china size 560 × 455mm complete with brackets, chromium-plated waste, overflow with chain and plastic plug						
white	nr	2.25	34.88	71.20	15.91	121.99
coloured	nr	2.25	34.88	82.41	17.59	134.88
Wash basin, vitreous china size 610 × 470mm complete with brackets, chromium-plated waste, overflow with chain and plastic plug						
white	nr	2.30	35.65	75.14	16.62	127.41
coloured	nr	2.30	35.65	82.43	17.71	135.79
Sinks and tops						
Stainless steel single sit-on type sink, complete with inset chromium-plated waste, overflow with chain and plastic plug, size						
1000 × 500mm	nr	1.30	20.15	94.00	17.12	131.27
1000 × 600mm	nr	1.40	21.70	100.98	18.40	141.08
1200 × 600mm	nr	1.50	23.25	113.59	20.53	157.37
Stainless steel single roll-edge type sink, complete with inset chromium-plated waste, overflow with chain and plastic plug, size						
1000 × 500mm	nr	1.30	20.15	97.06	17.58	134.79
1000 × 600mm	nr	1.40	21.70	105.30	19.05	146.05
1200 × 600mm	nr	1.50	23.25	117.76	21.15	162.16
Stainless steel double sit-on type sink, complete with inset chromium-plated waste, overflow with chain and plastic plug, size						
1500 × 500mm	nr	1.80	27.90	117.21	21.77	166.88
1500 × 600mm	nr	1.90	29.45	134.76	24.63	188.84

	Unit	Labour hours	Hours cost £	Materials £	O & P £	Total £
Stainless steel double roll-edge type sink, complete with inset chromium-plated waste, overflow with chain and plastic plug, size						
1000 × 500mm	nr	1.80	27.90	126.12	23.10	177.12
1000 × 600mm	nr	1.90	29.45	140.18	25.44	195.07
Belfast pattern white fireclay sink, complete with chromium-plated waste,chain and plastic plug, wall-mounted on brackets, size						
450 × 380 × 205mm	nr	2.35	36.43	125.97	24.36	186.75
610 × 455 × 255mm	nr	2.40	37.20	175.74	31.94	244.88
760 × 455 × 255mm	nr	2.45	37.98	236.05	41.10	315.13
WC suites						
Vitreous china low-level WC suite comprising pan, plastic seat and cover, 9 litre cistern, low-pressure ball valve, connecting pipework screwed to floor	nr	2.35	36.43	222.97	38.91	298.30
Vitreous china low-level close coupled WC suite comprising pan, plastic seat and cover, 9 litre cistern, low-pressure ball valve, connecting pipework screwed to floor	nr	2.45	37.98	299.71	50.65	388.34
Bidets						
Free-standing plain rim, vitreous china bidet excluding fittings						
white	nr	2.50	38.75	282.08	48.12	368.95
coloured	nr	2.50	38.75	303.45	51.33	393.53

	Unit	Labour hours	Hours cost £	Materials £	O & P £	Total £
Urinals						
Bowl urinals, white vitreous china, wall-mounted on hangers chromium-plated dome outlet						
430mm wide × 305mm high	nr	1.00	15.50	120.05	20.33	155.88
Stainless steel flush pipes and spreaders for bowl urinals, face fixed to wall						
one bowl set	nr	1.20	18.60	64.27	12.43	95.30
two bowl set	nr	1.75	27.13	111.68	20.82	159.63
three bowl set	nr	2.10	32.55	162.32	29.23	224.10
Automatic flushing cistern and fittings, white vitreous china, wall-mounted on brackets						
4.5 litre	nr	1.00	15.50	135.98	22.72	174.20
9 litre	nr	1.05	16.28	152.89	25.37	194.54
13.5 litre	nr	1.15	17.83	184.94	30.41	233.18
Modular slab urinal, white fireclay china, comprising back slabs without divisions, waterway channel, automatic flushing cistern, stainless steel flush pipes						
two persons	nr	5.20	80.60	656.47	110.56	847.63
three persons	nr	6.00	93.00	1076.57	175.44	1345.01
four persons	nr	7.00	108.50	1363.23	220.76	1692.49
Taps						
Chromium-plated pillar taps						
13mm	pr	0.40	6.20	26.02	4.83	37.05
19mm	pr	0.45	6.98	34.79	6.26	48.03

	Unit	Labour hours	Hours cost £	Materials £	O & P £	Total £
Chromium-plated mixer taps, 19mm						
cross-top handles	pr	0.60	9.30	64.23	11.03	84.56
lever handles	pr	0.60	9.30	86.52	14.37	110.19
Chromium-plated bidet monoblock and spray	nr	0.70	10.85	185.58	29.46	225.89
Valves						
Chromium-plated thermostatic exposed shower valve with flexible hose, slide rail and spray set	nr	1.40	21.70	219.70	36.21	277.61
Instant electric 9.5kw shower with flexible hose, slide rail and spray set	nr	1.50	23.25	182.33	30.84	236.42
Showers						
Shower cubicle size 788 × 842 × 2115mm in anodised aluminium frame and safety glass	nr	2.00	31.00	751.53	117.38	899.91
White glazed fireclay shower tray size 900 × 900 × 180mm with chromium-plated waste fitting	nr	1.80	27.90	194.79	33.40	256.09
White acrylic fireclay shower tray size 750 × 750 × 180mm with chromium-plated waste fitting	nr	1.30	20.15	102.84	18.45	141.44

	Unit	Labour hours	Hours cost £	Materials £	O & P £	Total £
P31 HOLES, CHASES, COVERS AND SUPPORTS FOR SERVICES						
Cutting holes for pipes up to 25mm diameter in walls 102.5mm thick						
commons	nr	0.40	6.20	0.00	0.93	7.13
facings	nr	0.70	10.85	0.00	1.63	12.48
engineering class A	nr	0.90	13.95	0.00	2.09	16.04
engineering class B	nr	0.75	11.63	0.00	1.74	13.37
Cutting holes for pipes up to 25mm diameter in walls 100mm thick						
concrete blocks	nr	0.30	4.65	0.00	0.70	5.35
Thermalite blocks	nr	0.20	3.10	0.00	0.47	3.57
aerated concrete blocks	nr	0.25	3.88	0.00	0.58	4.46
Cutting holes for pipes up to 25mm diameter in walls 225mm thick						
commons	nr	0.80	12.40	0.00	1.86	14.26
facings	nr	0.85	13.18	0.00	1.98	15.15
engineering class A	nr	1.30	20.15	0.00	3.02	23.17
engineering class B	nr	1.50	23.25	0.00	3.49	26.74
Cutting holes for pipes up to 25mm diameter in walls 100mm thick						
concrete blocks	nr	0.45	6.98	0.00	1.05	8.02
Thermalite blocks	nr	0.30	4.65	0.00	0.70	5.35
aerated concrete blocks	nr	0.35	5.43	0.00	0.81	6.24

	Unit	Labour hours	Hours cost £	Materials £	O & P £	Total £
Cutting holes for pipes up to 25 to 50mm diameter in walls 102.5mm thick						
commons	nr	0.50	7.75	0.00	1.16	8.91
facings	nr	0.80	12.40	0.00	1.86	14.26
engineering class A	nr	1.00	15.50	0.00	2.33	17.83
engineering class B	nr	0.85	13.18	0.00	1.98	15.15
Cutting holes for pipes up to 25 to 50mm diameter in walls 100mm thick						
concrete blocks	nr	0.40	6.20	0.00	0.93	7.13
Thermalite blocks	nr	0.30	4.65	0.00	0.70	5.35
aerated concrete blocks	nr	0.35	5.43	0.00	0.81	6.24
Cutting holes for pipes up to 25 to 50mm diameter in walls 225mm thick						
commons	nr	1.05	16.28	0.00	2.44	18.72
facings	nr	1.10	17.05	0.00	2.56	19.61
engineering class A	nr	1.60	24.80	0.00	3.72	28.52
engineering class B	nr	1.80	27.90	0.00	4.19	32.09
Cutting holes for pipes up to 25 to 50mm diameter in walls 100mm thick						
concrete blocks	nr	0.60	9.30	0.00	1.40	10.70
Thermalite blocks	nr	0.40	6.20	0.00	0.93	7.13
aerated concrete blocks	nr	0.45	6.98	0.00	1.05	8.02
Cutting and pinning ends of pipe support brackets and make good to						
concrete	nr	0.45	6.98	0.00	1.05	8.02
brickwork	nr	0.35	5.43	0.00	0.81	6.24
blockwork	nr	0.30	4.65	0.00	0.70	5.35
tiled walls	nr	0.50	7.75	0.00	1.16	8.91

	Unit	Labour hours	Hours cost £	Materials £	O & P £	Total £
Cut opening through cavity wall comprising facing bricks and concrete blocks						
balanced flue outlet	nr	0.90	13.95	0.00	2.09	16.04
150mm diameter flue pipe	nr	0.75	11.63	0.00	1.74	13.37
19mm overflow pipe	nr	0.20	3.10	0.00	0.47	3.57
Cutting chases for one pipe up to 25mm diameter in						
commons	nr	0.35	5.43	0.00	0.81	6.24
facings	nr	0.38	5.89	0.00	0.88	6.77
engineering class A	nr	0.45	6.98	0.00	1.05	8.02
engineering class B	nr	0.40	6.20	0.00	0.93	7.13
concrete blocks	nr	0.35	5.43	0.00	0.81	6.24
Thermalite blocks	nr	0.25	3.88	0.00	0.58	4.46
aerated concrete blocks	nr	0.20	3.10	0.00	0.47	3.57
Cutting chases for two pipes up to 25mm diameter or one pipe 50mm diameter in						
commons	m	0.45	6.98	0.00	1.05	8.02
facings	m	0.50	7.75	0.00	1.16	8.91
engineering class A	m	0.65	10.08	0.00	1.51	11.59
engineering class B	m	0.55	8.53	0.00	1.28	9.80
concrete blocks	m	0.40	6.20	0.00	0.93	7.13
Thermalite blocks	m	0.30	4.65	0.00	0.70	5.35
aerated concrete blocks	m	0.90	13.95	0.00	2.09	16.04
Make good surfaces both sides of chases						
plastered surfaces	m	0.15	2.33	0.70	0.45	3.48
tiled surfaces	m	0.25	3.88	2.33	0.93	7.14
concrete floor	m	0.35	5.43	0.70	0.92	7.04
granolithic floor	m	0.50	7.75	0.75	1.28	9.78
concrete soffit	m	0.55	8.53	0.70	1.38	10.61

	Unit	Labour hours	Hours cost £	Materials £	O & P £	Total £
R10 RAINWATER PIPEWORK/GUTTERS						
PVC-U rainwater pipe plugged to brickwork with pipe brackets and fitting clips at 2m maximum centres						
68mm diameter pipe	m	0.25	3.88	6.08	1.49	11.45
extra over for						
bend, 87.5 degrees	nr	0.25	3.88	5.84	1.46	11.17
offset	nr	0.25	3.88	3.48	1.10	8.46
branch	nr	0.25	3.88	11.64	2.33	17.84
shoe	nr	0.25	3.88	5.04	1.34	10.25
access pipe	nr	0.25	3.88	16.02	2.98	22.88
hopper head	nr	0.25	3.88	18.72	3.39	25.98
68mm square pipe	m	0.25	3.88	7.29	1.67	12.84
extra over for						
bend, 87.5 degrees	nr	0.25	3.88	3.73	1.14	8.75
offset	nr	0.25	3.88	6.55	1.56	11.99
branch	nr	0.25	3.88	13.19	2.56	19.62
shoe	nr	0.25	3.88	4.28	1.22	9.38
access pipe	nr	0.25	3.88	20.00	3.58	27.46
hopper head	nr	0.25	3.88	18.72	3.39	25.98
Aluminium rainwater pipe, straight, plain eared, spigot and socket dry joints plugged to brickwork						
63mm diameter pipe	m	0.28	4.34	16.63	3.15	24.12
extra over for						
offset, 75mm	nr	0.28	4.34	31.47	5.37	41.18
offset, 100mm	nr	0.28	4.34	31.47	5.37	41.18
offset, 150mm	nr	0.28	4.34	31.47	5.37	41.18
offset, 225mm	nr	0.28	4.34	37.77	6.32	48.43
offset, 300mm	nr	0.28	4.34	37.77	6.32	48.43

	Unit	Labour hours	Hours cost £	Materials £	O & P £	Total £
63mm diameter pipe (cont'd)						
bend, 92.5 degrees	nr	0.28	4.34	25.61	4.49	34.44
branch	nr	0.28	4.34	27.40	4.76	36.50
shoe	nr	0.28	4.34	9.54	2.08	15.96
standard hopper	nr	0.28	4.34	26.61	4.64	35.59
flatback hopper	nr	0.28	4.34	18.59	3.44	26.37
76mm diameter pipe	m	0.30	4.65	20.49	3.77	28.91
extra over for						
offset, 75mm	nr	0.30	4.65	38.72	6.51	49.88
offset, 100mm	nr	0.30	4.65	38.72	6.51	49.88
offset, 150mm	nr	0.30	4.65	38.72	6.51	49.88
offset, 225mm	nr	0.30	4.65	45.86	7.58	58.09
offset, 300mm	nr	0.30	4.65	46.85	7.73	59.23
bend, 92.5 degrees	nr	0.30	4.65	28.79	5.02	38.46
branch	nr	0.30	4.65	32.92	5.64	43.21
shoe	nr	0.30	4.65	13.19	2.68	20.52
standard hopper	nr	0.30	4.65	27.73	4.86	37.24
flatback hopper	nr	0.30	4.65	18.59	3.49	26.73
102mm diameter pipe	m	0.32	4.96	28.62	5.04	38.62
extra over for						
offset, 75mm	nr	0.32	4.96	47.07	7.80	59.83
offset, 100mm	nr	0.32	4.96	47.07	7.80	59.83
offset, 150mm	nr	0.32	4.96	47.07	7.80	59.83
offset, 225mm	nr	0.32	4.96	54.93	8.98	68.87
offset, 300mm	nr	0.32	4.96	54.93	8.98	68.87
bend, 92.5 degrees	nr	0.32	4.96	41.28	6.94	53.18
branch	nr	0.32	4.96	44.27	7.38	56.61
shoe	nr	0.32	4.96	17.41	3.36	25.73
standard hopper	nr	0.32	4.96	37.26	6.33	48.55
ornamental hopper	nr	0.32	4.96	130.20	20.27	155.43

	Unit	Labour hours	Hours cost £	Materials £	O & P £	Total £
Cast iron rainwater pipe, straight, ears cast on, spigot and socket dry joints plugged to brickwork						
65mm diameter pipe	m	0.25	3.88	21.20	3.76	28.84
extra over for						
offset, 75mm	nr	0.25	3.88	16.57	3.07	23.51
offset, 115mm	nr	0.25	3.88	16.57	3.07	23.51
offset, 150mm	nr	0.25	3.88	16.57	3.07	23.51
offset, 225mm	nr	0.25	3.88	19.30	3.48	26.65
offset, 305mm	nr	0.25	3.88	22.60	3.97	30.45
bend	nr	0.25	3.88	10.82	2.20	16.90
branch	nr	0.25	3.88	21.28	3.77	28.93
shoe	nr	0.25	3.88	15.34	2.88	22.10
eared shoe	nr	0.25	3.88	17.68	3.23	24.79
flat hopper	nr	0.25	3.88	13.82	2.65	20.35
rectangular hopper	nr	0.25	3.88	61.66	9.83	75.37
75mm diameter pipe	m	0.30	4.65	22.25	4.04	30.94
extra over for						
offset, 75mm	nr	0.30	4.65	16.57	3.18	24.40
offset, 115mm	nr	0.30	4.65	16.57	3.18	24.40
offset, 150mm	nr	0.30	4.65	16.57	3.18	24.40
offset, 225mm	nr	0.30	4.65	19.30	3.59	27.54
offset, 305mm	nr	0.30	4.65	23.71	4.25	32.61
bend	nr	0.30	4.65	13.15	2.67	20.47
branch	nr	0.30	4.65	23.02	4.15	31.82
shoe	nr	0.30	4.65	15.34	3.00	22.99
eared shoe	nr	0.30	4.65	17.68	3.35	25.68
flat hopper	nr	0.30	4.65	15.69	3.05	23.39
rectangular hopper	nr	0.30	4.65	61.56	9.93	76.14
100mm diameter pipe	m	0.35	5.43	29.07	5.17	39.67
extra over for						
offset, 75mm	nr	0.35	5.43	31.28	5.51	42.21
offset, 115mm	nr	0.35	5.43	31.88	5.60	42.90
offset, 150mm	nr	0.35	5.43	31.28	5.51	42.21
offset, 225mm	nr	0.35	5.43	37.87	6.49	49.79
offset, 305mm	nr	0.35	5.43	61.56	10.05	77.03
bend	nr	0.35	5.43	18.57	3.60	27.59

	Unit	Labour hours	Hours cost £	Materials £	O & P £	Total £
100mm diameter pipe (cont'd)						
branch	nr	0.35	5.43	27.35	4.92	37.69
shoe	nr	0.35	5.43	20.68	3.92	30.02
eared shoe	nr	0.35	5.43	23.50	4.34	33.26
flat hopper	nr	0.35	5.43	34.78	6.03	46.24
rectangular hopper	nr	0.35	5.43	61.56	10.05	77.03
PVC-U half round rainwater gutter, fixed to timber with support brackets at 1m maximum centres						
76mm elliptical gutter	m	0.20	3.10	6.58	1.45	11.13
extra over for						
running outlet	nr	0.20	3.10	6.00	1.37	10.47
angle	nr	0.20	3.10	6.54	1.45	11.09
stop end outlet	nr	0.10	1.55	5.26	1.02	7.83
stop end	nr	0.10	1.55	3.22	0.72	5.49
112mm half-round gutter	m	0.26	4.03	4.86	1.33	10.22
extra over for						
running outlet	nr	0.26	4.03	4.23	1.24	9.50
angle	nr	0.26	4.03	4.82	1.33	10.18
stop end outlet	nr	0.13	2.02	4.23	0.94	7.18
stop end	nr	0.13	2.02	2.17	0.63	4.81
150mm half-round gutter	m	0.30	4.65	7.17	1.77	13.59
extra over for						
running outlet	nr	0.30	4.65	9.76	2.16	16.57
angle	nr	0.30	4.65	4.81	1.42	10.88
stop end outlet	nr	0.15	2.33	6.14	1.27	9.73
stop end	nr	0.15	2.33	1.97	0.64	4.94

	Unit	Labour hours	Hours cost £	Materials £	O & P £	Total £
Aluminium half round rainwater gutter with mastic joints, fixed to timber with support brackets at 1m maximum centres						
100mm gutter	m	0.30	4.35	13.33	2.65	20.33
extra over for						
running outlet	nr	0.30	4.35	11.89	2.44	18.68
angle	nr	0.30	4.35	6.99	1.70	13.04
stop end outlet	nr	0.15	2.18	6.83	1.35	10.36
stop end	nr	0.15	2.18	2.63	0.72	5.53
125mm gutter	m	0.32	4.64	16.80	3.22	24.66
extra over for						
running outlet	nr	0.32	4.64	11.37	2.40	18.41
angle	nr	0.32	4.64	9.53	2.13	16.30
stop end outlet	nr	0.16	2.32	7.86	1.53	11.71
stop end	nr	0.16	2.32	3.42	0.86	6.60
Aluminium ogee rainwater gutter with mastic joints, fixed to timber with support brackets at 1m maximum centres						
120 × 75mm gutter	m	0.30	4.35	17.22	3.24	24.81
extra over for						
running outlet	nr	0.30	4.35	21.70	3.91	29.96
angle	nr	0.30	4.35	18.17	3.38	25.90
stop end outlet	nr	0.15	2.18	30.32	4.87	37.37
stop end	nr	0.15	2.18	4.23	0.96	7.37
155 × 100mm gutter	m	0.32	4.64	32.49	5.57	42.70
extra over for						
running outlet	nr	0.32	4.64	28.69	5.00	38.33
angle	nr	0.32	4.64	28.05	4.90	37.59
stop end outlet	nr	0.16	2.32	23.38	3.86	29.56
stop end	nr	0.16	2.32	17.07	2.91	22.30

	Unit	Labour hours	Hours cost £	Materials £	O & P £	Total £
Cast iron half-round rainwater gutter with mastic joints, primed, fixed to timber with support brackets at 1m maximum centres						
100mm gutter	m	0.36	5.22	11.77	2.55	19.54
extra over for						
running outlet	nr	0.36	5.22	7.56	1.92	14.70
angle	nr	0.36	5.22	7.76	1.95	14.93
stop end outlet	nr	0.18	2.61	8.87	1.72	13.20
stop end	nr	0.18	2.61	2.61	0.78	6.00
115mm gutter	m	0.38	5.51	12.15	2.65	20.31
extra over for						
running outlet	nr	0.38	5.51	8.24	2.06	15.81
angle	nr	0.38	5.51	7.99	2.03	15.53
stop end outlet	nr	0.19	2.76	9.94	1.90	14.60
stop end	nr	0.19	2.76	3.38	0.92	7.06
125mm gutter	m	0.40	5.80	14.05	2.98	22.83
extra over for						
running outlet	nr	0.40	5.80	7.84	2.05	15.69
angle	nr	0.40	5.80	9.42	2.28	17.50
stop end outlet	nr	0.20	2.90	11.34	2.14	16.38
stop end	nr	0.20	2.90	3.38	0.94	7.22
150mm gutter	m	0.44	6.38	23.47	4.48	34.33
extra over for						
running outlet	nr	0.44	6.38	14.71	3.16	24.25
angle	nr	0.44	6.38	17.21	3.54	27.13
stop end outlet	nr	0.22	3.19	18.66	3.28	25.13
stop end	nr	0.22	3.19	4.68	1.18	9.05

	Unit	Labour hours	Hours cost £	Materials £	O & P £	Total £
Roof outlets						
Cast iron circular roof outlet with flat grate, diameter						
50mm	nr	0.50	7.25	81.01	13.24	101.50
75mm	nr	0.60	8.70	88.35	14.56	111.61
100mm	nr	0.70	10.15	103.44	17.04	130.63
Aluminium circular roof outlet with domed grate, diameter						
50mm	nr	0.50	7.25	74.70	12.29	94.24
75mm	nr	0.60	8.70	85.28	14.10	108.08
100mm	nr	0.70	10.15	110.08	18.03	138.26
150mm	nr	0.80	11.60	123.20	20.22	155.02
Plastic wire balloon guard for pipes and outlets, diameter						
50mm	nr	0.05	0.73	2.14	0.43	3.29
63mm	nr	0.05	0.73	2.27	0.45	3.44
75mm	nr	0.05	0.73	2.34	0.46	3.52
100mm	nr	0.05	0.73	2.36	0.46	3.55
Galvanised wire balloon guard for pipes and outlets, diameter						
50mm	nr	0.05	0.73	2.79	0.53	4.04
63mm	nr	0.05	0.73	1.85	0.39	2.96
75mm	nr	0.05	0.73	1.89	0.39	3.01
100mm	nr	0.05	0.73	2.36	0.46	3.55
Copper wire balloon guard for pipes and outlets, diameter						
50mm	nr	0.05	0.73	2.13	0.11	0.83
63mm	nr	0.05	0.73	2.17	0.43	3.33
75mm	nr	0.05	0.73	2.33	0.46	3.51
100mm	nr	0.05	0.73	3.27	0.60	4.59

	Unit	Labour hours	Hours cost £	Materials £	O & P £	Total £
R11 DRAINAGE ABOVE GROUND						
MuPVC waste system, solvent welded joints, clips at 500mm maximum centres, plugged to brickwork						
32mm diameter pipe	m	0.25	3.88	3.29	1.07	8.24
extra over for						
bend, 45°	nr	0.24	3.72	2.00	0.86	6.58
bend, 87.5°	nr	0.24	3.72	2.00	0.86	6.58
long tail bend, 90°	nr	0.24	3.72	2.00	0.86	6.58
tee	nr	0.24	3.72	3.42	1.07	8.21
bottle trap	nr	0.24	3.72	4.65	1.26	9.63
tubular P trap	nr	0.24	3.72	4.57	1.24	9.53
tubular S trap	nr	0.24	3.72	5.48	1.38	10.58
40mm diameter pipe	m	0.28	4.34	3.54	1.18	9.06
extra over for						
bend, 45°	nr	0.26	4.03	2.29	0.95	7.27
bend, 87.5°	nr	0.26	4.03	2.29	0.95	7.27
long tail bend, 90°	nr	0.26	4.03	2.29	0.95	7.27
tee	nr	0.26	4.03	4.12	1.22	9.37
cross tee	nr	0.26	4.03	11.03	2.26	17.32
bottle trap	nr	0.26	4.03	5.28	1.40	10.71
tubular P trap	nr	0.26	4.03	5.48	1.43	10.94
tubular S trap	nr	0.26	4.03	6.52	1.58	12.13
running P trap	nr	0.26	4.03	7.84	1.78	13.65
50mm diameter pipe	m	0.34	5.27	4.98	1.54	11.79
extra over for						
bend, 45°	nr	0.32	4.96	3.21	1.23	9.40
bend, 87.5°	nr	0.32	4.96	3.21	1.23	9.40
long tail bend, 90°	nr	0.32	4.96	3.21	1.23	9.40
tee	nr	0.32	4.96	6.91	1.78	13.65
cross tee	nr	0.32	4.96	12.79	2.66	20.41

	Unit	Labour hours	Hours cost £	Materials £	O & P £	Total £
MuPVC waste system, push-fit joints, clips at 500mm maximum centres, plugged to brickwork						
32mm diameter pipe	m	0.20	3.10	1.18	0.64	4.92
extra over for						
bend, 45°	nr	0.18	2.79	1.00	0.57	4.36
knuckle bend, 90°	nr	0.18	2.79	0.98	0.57	4.34
spigot bend, 90°	nr	0.18	2.79	1.23	0.60	4.62
tee	nr	0.18	2.79	1.56	0.65	5.00
P trap, 38mm	nr	0.18	2.79	4.57	1.10	8.46
bottle trap, 38mm	nr	0.18	2.79	4.65	1.12	8.56
anti-syphon bottle trap, 76mm	nr	0.18	2.79	7.43	1.53	11.75
connection to back inlet gulley, caulking bush	nr	0.16	2.48	3.12	0.84	6.44
40mm diameter pipe	m	0.24	3.72	2.00	0.86	6.58
extra over for						
bend, 45°	nr	0.20	3.10	1.06	0.62	4.78
knuckle bend, 90°	nr	0.20	3.10	1.00	0.62	4.72
spigot bend, 90°	nr	0.20	3.10	1.29	0.66	5.05
tee	nr	0.20	3.10	1.61	0.71	5.42
P trap, 38mm	nr	0.20	3.10	5.48	1.29	9.87
bottle trap, 38mm	nr	0.20	3.10	5.28	1.26	9.64
anti-syphon bottle trap, 76mm	nr	0.20	3.10	8.60	1.76	13.46
connection to back inlet gulley, caulking bush	nr	0.18	2.79	2.29	0.76	5.84
PVC-U soil system, solvent-welded joints, holderbats at 1250mm maximum centres, plugged to brickwork						
82mm diameter pipe	m	0.34	5.27	13.49	2.81	21.57
extra over for						
bend, 92.5°	nr	0.30	4.65	13.60	2.74	20.99
bend, 135°	nr	0.30	4.65	13.60	2.74	20.99

	Unit	Labour hours	Hours cost £	Materials £	O & P £	Total £
82mm diameter pipe (cont'd)						
bend, spigot/spigot	nr	0.30	4.65	13.60	2.74	20.99
branch, single, 92.5°	nr	0.38	5.89	20.00	3.88	29.77
branch, single, 135°	nr	0.38	5.89	20.00	3.88	29.77
access pipe, single branch	nr	0.38	5.89	20.00	3.88	29.77
access cap	nr	0.22	3.41	12.27	2.35	18.03
vent cowl	nr	0.22	3.41	3.68	1.06	8.15
110mm diameter pipe	m	0.38	5.89	11.53	2.61	20.03
extra over for						
bend, 92.5°	nr	0.34	5.27	16.15	3.21	24.63
bend, 135°	nr	0.34	5.27	16.15	3.21	24.63
bend, variable	nr	0.34	5.27	16.15	3.21	24.63
bend, spigot/spigot	nr	0.34	5.27	14.82	3.01	23.10
branch, single, 92.5°	nr	0.40	6.20	21.04	4.09	31.33
branch, single, 135°	nr	0.40	6.20	21.04	4.09	31.33
branch, single, 92.5° spigot outlet	nr	0.40	6.20	21.04	4.09	31.33
access pipe connector	nr	0.34	5.27	21.58	4.03	30.88
access pipe, single branch	nr	0.34	5.27	21.04	3.95	30.26
access cap	nr	0.38	5.89	12.96	2.83	21.68
WC manifold connector	nr	0.38	5.89	9.88	2.37	18.14
vent cowl	nr	0.22	3.41	3.88	1.09	8.38
150mm diameter pipe	m	0.42	6.51	31.32	5.67	43.50
extra over for						
bend, 92.5°	nr	0.38	5.89	43.88	7.47	57.24
bend, 135°	nr	0.38	5.89	43.88	7.47	57.24
bend, spigot/spigot	nr	0.38	5.89	43.88	7.47	57.24
bend, spigot/socket	nr	0.38	5.89	43.88	7.47	57.24
branch, single, 92.5°	nr	0.42	6.51	72.27	11.82	90.60
branch, single, 135°	nr	0.42	6.51	72.27	11.82	90.60
access cap	nr	0.40	6.20	35.55	6.26	48.01
vent cowl	nr	0.24	3.72	10.94	2.20	16.86

	Unit	Labour hours	Hours cost £	Materials £	O & P £	Total £
Cast iron soil system with flexible joints, pipe brackets at 2m maximum centres, plugged to brickwork						
50mm diameter pipe	m	0.55	8.53	21.66	4.53	34.71
extra over for						
bend, short radius	nr	0.45	6.98	10.72	2.65	20.35
bend, short radius with door	nr	0.45	6.98	26.44	5.01	38.43
branch, single, plain	nr	0.50	7.75	16.14	3.58	27.47
branch, single with door	nr	0.50	7.75	31.86	5.94	45.55
branch, double, plain	nr	0.50	7.75	28.39	5.42	41.56
P trap with door	nr	0.45	6.98	34.67	6.25	47.89
75mm diameter pipe	m	0.60	9.30	21.44	4.61	35.35
extra over for						
bend, short radius	nr	0.50	7.75	10.72	2.77	21.24
bend, short radius with door	nr	0.50	7.75	26.44	5.13	39.32
bend, long radius	nr	0.50	7.75	21.26	4.35	33.36
bend, long radius, access	nr	0.50	7.75	26.44	5.13	39.32
branch, single, plain	nr	0.55	8.53	16.14	3.70	28.36
branch, single with door	nr	0.55	8.53	31.86	6.06	46.44
branch, double, plain	nr	0.55	8.53	27.15	5.35	41.03
offset, 75mm projection	nr	0.50	7.75	10.59	2.75	21.09
offset, 115mm projection	nr	0.50	7.75	13.19	3.14	24.08
offset, 150mm projection	nr	0.50	7.75	13.19	3.14	24.08
offset, 225mm projection	nr	0.50	7.75	16.53	3.64	27.92
offset, 300mm projection	nr	0.50	7.75	19.50	4.09	31.34
blank end, plain	nr	0.50	7.75	3.85	1.74	13.34
blank end, drilled and tapped	nr	0.50	7.75	9.15	2.54	19.44
P trap, plain	nr	0.50	7.75	20.68	4.26	32.69
P trap with door	nr	0.50	7.75	36.40	6.62	50.77

	Unit	Labour hours	Hours cost £	Materials £	O & P £	Total £
WC connector, 305mm effective length	nr	0.50	7.75	19.80	4.13	31.68
100mm diameter pipe	m	0.65	10.08	32.12	6.33	48.52
extra over for						
bend, short radius	nr	0.55	8.53	14.84	3.50	26.87
bend, short radius with door	nr	0.55	8.53	31.39	5.99	45.90
bend, long radius	nr	0.55	8.53	24.04	4.88	37.45
bend, long radius, access	nr	0.55	8.53	40.61	7.37	56.51
bend, long radius, heel rest	nr	0.55	8.53	29.11	5.65	43.28
bend, long tail, 87.5°	nr	0.55	8.53	19.18	4.16	31.86
branch, single, plain	nr	0.60	9.30	22.94	4.84	37.08
branch, single with door	nr	0.60	9.30	39.53	7.32	56.15
branch, double, plain	nr	0.60	9.30	28.39	5.65	43.34
branch, double with door	nr	0.60	9.30	44.95	8.14	62.39
branch, corner	nr	0.60	9.30	37.35	7.00	53.65
offset, 75mm projection	nr	0.55	8.53	15.61	3.62	27.76
offset, 115mm projection	nr	0.55	8.53	18.61	4.07	31.21
offset, 150mm projection	nr	0.55	8.53	18.61	4.07	31.21
offset, 225mm projection	nr	0.55	8.53	21.34	4.48	34.34
offset, 300mm projection	nr	0.55	8.53	24.04	4.88	37.45
blank end, plain	nr	0.55	8.53	4.52	1.96	15.00
blank end, drilled and tapped	nr	0.55	8.53	9.83	2.75	21.11
P trap, plain	nr	0.55	8.53	23.79	4.85	37.16
P trap with door	nr	0.55	8.53	40.36	7.33	56.22
WC connector, 305mm effective length	nr	0.55	8.53	13.50	3.30	25.33

	Unit	Labour hours	Hours cost £	Materials £	O & P £	Total £
150mm diameter pipe	m	0.70	10.85	62.38	10.98	84.21
extra over for						
bend, short radius	nr	0.60	9.30	26.51	5.37	41.18
bend, short radius with door	nr	0.60	9.30	44.60	8.09	61.99
bend, long radius	nr	0.60	9.30	43.53	7.92	60.75
bend, long radius, access	nr	0.60	9.30	6.15	2.32	17.77
branch, single, plain	nr	0.65	10.08	47.20	8.59	65.87
branch, single with door	nr	0.65	10.08	65.34	11.31	86.73
branch, double, plain	nr	0.65	10.08	70.13	12.03	92.24
blank end, plain	nr	0.60	9.30	6.52	2.37	18.19
blank end, drilled and tapped	nr	0.60	9.30	118.81	19.22	147.33
P trap with door	nr	0.60	9.30	70.28	11.94	91.52
MuPVC overflow system, solvent-welded joints, clips at 500mm maximum centres, plugged to brickwork						
19mm diameter pipe	m	0.20	3.10	1.87	0.75	5.72
extra over for						
bend	nr	0.18	2.79	1.37	0.62	4.78
tee	nr	0.18	2.79	1.42	0.63	4.84
connector, bent tank	nr	0.20	3.10	2.03	0.77	5.90
Polypropylene traps, screwed joints to outlet and pipe						
Bottle trap, 38mm seal						
32mm	nr	0.30	4.65	4.65	1.40	10.70
40mm	nr	0.35	5.43	4.74	1.52	11.69
Bottle trap, 76mm seal						
32mm	nr	0.30	4.65	4.65	1.40	10.70
40mm	nr	0.35	5.43	4.74	1.52	11.69
Anti-syphon bottle trap						
32mm	nr	0.30	4.65	7.43	1.81	13.89
40mm	nr	0.35	5.43	7.72	1.97	15.12
Tubular S trap						
32mm	nr	0.30	4.65	5.48	1.52	11.65
40mm	nr	0.35	5.43	6.52	1.79	13.74
Tubular P trap						
32mm	nr	0.30	4.65	4.57	1.38	10.60
40mm	nr	0.35	5.43	5.48	1.64	12.54

	Unit	Labour hours	Hours cost £	Materials £	O & P £	Total £
Running P trap						
40mm	nr	0.35	5.43	7.84	1.99	15.25
Bath trap						
40mm	nr	0.35	5.43	6.88	1.85	14.15
Bath trap with overflow						
40mm	nr	0.35	5.43	6.43	1.78	13.63
Washing machine, half trap						
40mm	nr	0.35	5.43	4.57	1.50	11.49

	Unit	Labour hours	Hours cost £	Materials £	O & P £	Total £
R12 DRAINAGE BELOW GROUND						
Timesaver cast iron drainage, jointed with flexible couplings and laid in trenches prepared by others						
100mm diameter pipe						
laid straight	m	0.40	6.20	33.00	5.88	45.08
laid vertically	m	0.75	11.63	33.00	6.69	51.32
in lengths less than 3m	m	0.70	10.85	36.06	7.04	53.95
Extra over for						
standard coupling	nr	0.15	2.33	19.87	3.33	25.52
continuity clip for standard coupling	nr	0.10	1.55	1.11	0.40	3.06
stepped coupling	nr	0.15	2.33	19.22	3.23	24.78
continuity clip for stepped coupling	nr	0.10	1.55	1.22	0.42	3.19
hanging bracket	nr	0.50	7.75	13.03	3.12	23.90
bend, 87.5°	nr	0.40	6.20	39.33	6.83	52.36
bend, 80°	nr	0.40	6.20	39.33	6.83	52.36
bend, 67°	nr	0.40	6.20	38.25	6.67	51.12
bend, 60°	nr	0.40	6.20	38.25	6.67	51.12
bend, 45°	nr	0.40	6.20	36.74	6.44	49.38
bend, 45°	nr	0.40	6.20	36.74	6.44	49.38
bend, 22.5°	nr	0.40	6.20	34.36	6.08	46.64
bend, 10°	nr	0.40	6.20	31.88	5.71	43.79
bend, 87.5°, heel rest	nr	0.40	6.20	35.69	6.28	48.17
bend, access rear,	nr	0.40	6.20	74.51	12.11	92.82
87.5°	nr	0.40	6.20	74.51	12.11	92.82
bend, access rear, 45°	nr	0.40	6.20	74.51	12.11	92.82
bend, access side, 87.5°	nr	0.40	6.20	81.03	13.08	100.31
bend, access side,	nr	0.40	6.20	81.03	13.08	100.31
45°	nr	0.40	6.20	81.03	13.08	100.31

	Unit	Labour hours	Hours cost £	Materials £	O & P £	Total £
bend, long radius, 87.5°	nr	0.40	6.20	53.70	8.99	68.89
bend, long radius, 87.5°, with heel rest	nr	0.40	6.20	55.80	9.30	71.30
bend, long tail, 87.5°	nr	0.40	6.20	73.04	11.89	91.13
bend, large radius, 87.5°	nr	0.40	6.20	77.96	12.62	96.78
branch, 100 × 100mm, 87.5°	nr	0.60	9.30	51.77	9.16	70.23
branch, 100 × 100mm, 67.5°	nr	0.60	9.30	53.44	9.41	72.15
branch, 100 × 100mm, 45°	nr	0.60	9.30	52.88	9.33	71.51
branch, 100 × 100mm, 87.5°, right-hand access	nr	0.60	9.30	88.52	14.67	112.49
branch, 100 × 100mm, 45°, right-hand access right-hand access	nr	0.60	9.30	88.52	14.67	112.49
branch, 100 × 100mm, 87.5°, rear access	nr	0.60	9.30	88.52	14.67	112.49
branch, 100 × 100mm, 45°, rear access	nr	0.60	9.30	88.52	14.67	112.49
branch, 100 × 100mm, 87.5°, left-hand access	nr	0.60	9.30	88.52	14.67	112.49
branch, 100 × 100mm, 45°, left-hand access	nr	0.60	9.30	88.52	14.67	112.49
branch, 100 × 100mm, 87.5°, double plain	nr	0.60	9.30	68.69	11.70	89.69
socket ferrule with cap	nr	0.40	6.20	34.47	6.10	46.77
blank cap plain	nr	0.40	6.20	15.33	3.23	24.76
cleaning arm bend	nr	0.60	9.30	55.72	9.75	74.77
transitional pipe, socket for clayware	nr	0.40	6.20	40.16	6.95	53.31
transitional pipe, socket for cast iron	nr	0.40	6.20	31.27	5.62	43.09

	Unit	Labour hours	Hours cost £	Materials £	O & P £	Total £
150mm diameter pipe						
laid straight	m	0.60	9.30	59.11	10.26	78.67
laid vertically	m	0.80	12.40	59.11	10.73	82.24
in lengths less than 3m	m	0.90	13.95	61.29	11.29	86.53
Extra over for						
standard coupling	nr	0.15	2.18	23.11	3.79	29.08
continuity clip for standard coupling	nr	0.10	1.45	1.31	0.41	3.17
stepped coupling	nr	0.15	2.33	1.43	0.56	4.32
continuity clip for stepped coupling	nr	0.10	1.55	1.49	0.46	3.50
hanging bracket	nr	0.50	7.75	16.09	3.58	27.42
bend, 87.5°	nr	0.70	10.85	77.42	13.24	101.51
bend, 80°	nr	0.70	10.85	68.19	11.86	90.90
bend, 45°	nr	0.70	10.85	58.41	10.39	79.65
bend, 35°	nr	0.70	10.85	58.41	10.39	79.65
bend, 22.5°	nr	0.70	10.85	56.05	10.04	76.94
bend, 10°	nr	0.70	10.85	52.05	9.44	72.34
bend, 87.5°, heel rest	nr	0.70	10.85	86.45	14.60	111.90
bend, access rear, 87.5°	nr	0.70	10.85	150.80	24.25	185.90
bend, access rear, 45°	nr	0.70	10.85	150.80	24.25	185.90
bend, access side, 87.5°	nr	0.70	10.85	150.80	24.25	185.90
bend, access side, 45°	nr	0.70	10.85	150.80	24.25	185.90
bend, long radius, 87.5°	nr	0.70	10.85	52.51	9.50	72.86
bend, long radius, 87.5°, with heel rest	nr	0.70	10.85	91.60	15.37	117.82
branch, 150 × 100mm, 87.5°	nr	0.90	13.95	97.69	16.75	128.39
branch, 150 × 100mm, 67.5°	nr	0.90	13.95	99.72	17.05	130.72
branch, 150 × 100mm, 45°	nr	0.90	13.95	94.67	16.29	124.91

	Unit	Labour hours	Hours cost £	Materials £	O & P £	Total £
branch, 150 × 100mm, 87.5°, right-hand access	nr	0.90	13.95	95.06	16.35	125.36
branch, 150 × 100mm, 45°, right-hand access	nr	0.90	13.95	95.06	16.35	125.36
branch, 150 × 100mm, 87.5°, rear access	nr	0.90	13.95	184.39	29.75	228.09
branch, 150 × 100mm, 45°, rear access	nr	0.90	13.95	184.39	29.75	228.09
branch, 150 × 100mm, 87.5°, left-hand access	nr	0.90	13.95	184.88	29.82	228.65
socket ferrule with cap	nr	0.50	7.75	62.21	10.49	80.45
blank cap plain	nr	0.50	7.75	29.69	5.62	43.06
transitional pipe, socket for clayware	nr	0.60	9.30	53.01	9.35	71.66
transitional pipe, socket for cast iron	nr	0.60	9.30	51.41	9.11	69.82
gulley trap plain, 150mm × 87.5°	nr	0.90	13.95	104.98	17.84	136.77
225mm diameter pipe						
laid straight	m	1.10	17.05	150.34	25.11	192.50
laid vertically	m	1.30	20.15	150.34	25.57	196.06
in lengths less than 3m	m	1.50	23.25	154.82	26.71	204.78
Extra over for						
standard coupling	nr	0.20	3.10	1.97	0.76	5.83
continuity clip for standard coupling	nr	0.15	2.33	1.76	0.61	4.70
bend, 87.5°	nr	1.00	15.50	209.12	33.69	258.31
bend, 45°	nr	1.00	15.50	209.12	33.69	258.31
bend, 22.5°	nr	1.00	15.50	209.12	33.69	258.31
bend, 10°	nr	1.00	15.50	209.12	33.69	258.31
bend, 87.5°, heel rest						
bend, access rear, 87.5°	nr	1.00	15.50	317.98	50.02	383.50

	Unit	Labour hours	Hours cost £	Materials £	O & P £	Total £
bend, access rear, 45°	nr	1.00	15.50	317.98	50.02	383.50
bend, access side, 87.5°	nr	1.00	15.50	317.98	50.02	383.50
bend, access side, 45°	nr	1.00	15.50	317.98	50.02	383.50
branch, 225 × 100mm, 87.5°	nr	1.20	18.60	261.50	42.02	322.12
branch, 225 × 100mm, 45°	nr	1.20	18.60	261.50	42.02	322.12
branch, 225 × 150mm, 87.5°	nr	1.20	18.60	273.72	43.85	336.17
branch, 225 × 150mm, 45°	nr	1.20	18.60	273.12	43.76	335.48
branch, 225 × 225mm, 87.5°	nr	1.20	18.60	297.18	47.37	363.15
branch, 225 × 225mm, 45°	nr	1.20	18.60	297.18	47.37	363.15
branch, 225 × 100mm, 45°, right-hand access	nr	1.20	18.60	380.87	59.92	459.39
branch, 225 × 150mm, 87.5°, right-hand access	nr	1.20	18.60	392.11	61.61	472.32
branch, 225 × 150mm, 45°, right-hand access	nr	1.20	18.60	392.11	61.61	472.32
branch, 225 × 225mm, 87.5°, right-hand access	nr	1.20	18.60	405.50	63.62	487.72
branch, 225 × 225mm, 45°, right-hand access	nr	1.20	18.60	405.50	63.62	487.72
branch, 225 × 100mm, 45°, rear access	nr	1.20	18.60	374.33	58.94	451.87
branch, 225 × 150mm, 87.5°, rear access	nr	1.20	18.60	374.33	58.94	451.87
branch, 225 × 150mm, 45°, rear access	nr	1.20	18.60	392.11	61.61	472.32

	Unit	Labour hours	Hours cost £	Materials £	O & P £	Total £
branch, 225 × 225mm, 87.5°, rear access	nr	1.20	18.60	405.50	63.62	487.72
branch, 225 × 225mm, 45°, rear access	nr	1.20	18.60	405.50	63.62	487.72
branch, 225 × 100mm, 87.5°, left-hand access	nr	1.20	18.60	374.33	58.94	451.87
branch, 225 × 100mm, 45°, left-hand access	nr	1.20	18.60	374.33	58.94	451.87
branch, 225 × 150mm, 87.5°, left-hand access	nr	1.20	18.60	392.11	61.61	472.32
branch, 225 × 225mm, 87.5°, left-hand access	nr	1.20	18.60	392.11	61.61	472.32
branch, 225 × 225mm, 45°, left-hand access	nr	1.20	18.60	405.50	63.62	487.72
diminishing pipe 225 to 100m	nr	0.70	10.85	99.53	16.56	126.94
diminishing pipe 225 to 150m	nr	0.70	10.85	99.53	16.56	126.94
socket ferrule with cap	nr	0.50	7.75	237.06	36.72	281.53
blank cap plain	nr	0.50	7.75	88.27	14.40	110.42
transitional pipe, socket for clayware	nr	0.60	9.30	149.77	23.86	182.93
transitional pipe, socket for cast iron	nr	0.60	9.30	158.91	25.23	193.44

Gullies and traps

	Unit	Labour hours	Hours cost £	Materials £	O & P £	Total £
Gulley trap plain, 100mm × 87.5°	nr	0.70	10.85	53.95	9.72	74.52
Gulley trap plain, 100mm × 45°	nr	0.70	10.85	60.12	10.65	81.62
Gulley trap plain, 150mm × 87.5°	nr	0.70	10.85	104.98	17.37	133.20
Gulley trap, 150mm × 87.5°, surface access	nr	0.70	10.85	99.33	16.53	126.71

	Unit	Labour hours	Hours cost £	Materials £	O & P £	Total £
Gulley trap, 150mm × 87.5°, 225mm inlet	nr	0.70	10.85	83.00	14.08	107.93
Gulley trap running, 150mm double access, 100 × 87.5°	nr	0.90	13.95	209.12	33.46	256.53
Gulley trap running, 150mm double access, 150 × 87.5°	nr	0.90	13.95	329.70	51.55	395.20
Gulley trap running, 150mm double access, 225 × 87.5°	nr	0.90	13.95	358.30	55.84	428.09
Gulley inlet plain	nr	0.90	13.95	51.49	9.82	75.26
Gulley inlet, 100 × 100mm with horizontal branch	nr	0.90	13.95	59.62	11.04	84.61
Gulley inlet, 100 × 100mm with vertical branch	nr	0.90	13.95	60.50	11.17	85.62
Raising piece, plain, 100 × 75mm	nr	0.40	6.20	42.09	7.24	55.53
Raising piece, plain, 100 × 150mm	nr	0.40	6.20	45.12	7.70	59.02
Raising piece, plain, 100 × 225mm	nr	0.40	6.20	47.51	8.06	61.77
Raising piece, plain, 100 × 300mm	nr	0.40	6.20	53.17	8.91	68.28
Notched grate, 200mm	nr	0.10	1.55	5.60	1.07	8.22
Plain grate, 200mm	nr	0.10	1.55	5.13	1.00	7.68
Solid cover, 200mm	nr	0.20	3.10	7.47	1.59	12.16
Hinged and locking grate, 200mm	nr	0.25	3.88	30.33	5.13	39.34
Sealed plate and frame, 200mm	nr	0.25	3.88	31.49	5.30	40.67
Trapless gulley, 87.5° × 100mm	nr	2.00	31.00	28.66	8.95	68.61
Dean's trap, 87.5° × 100mm with galvanised sediment pan	nr	3.00	46.50	141.58	28.21	216.29
Gulley trap 425mm deep, 87.5° × 100mm with galvanised sediment pan	nr	2.70	41.85	141.58	27.51	210.94

	Unit	Labour hours	Hours cost £	Materials £	O & P £	Total £
S10 HOT AND COLD WATER						
Copper pipe, lead-free pre-soldered capillary joints and fittings, clips at 1250mm maximum centres						
8mm diameter, to timber	m	0.20	3.10	1.29	0.66	5.05
8mm diameter, plugged and screwed	m	0.22	3.41	1.59	0.75	5.75
extra over for						
straight coupling	nr	0.18	2.79	1.08	0.58	4.45
reduced coupling, 8 × 6mm	nr	0.18	2.79	3.90	1.00	7.69
straight female connector, 8mm × 1/4in	nr	0.18	2.79	4.13	1.04	7.96
straight male connector, 8mm × 1/4in	nr	0.18	2.79	2.77	0.83	6.39
reducer, 8 × 6mm	nr	0.18	2.79	3.11	0.89	6.79
elbow, 8mm	nr	0.18	2.79	2.34	0.77	5.90
street elbow, 8mm	nr	0.18	2.79	3.19	0.90	6.88
equal tee, 8mm	nr	0.22	3.41	3.74	1.07	8.22
stop end, 8mm	nr	0.18	2.79	3.39	0.93	7.11
10mm diameter, to timber	m	0.20	3.10	1.51	0.69	5.30
10mm diameter, plugged and screwed	m	0.22	3.41	1.80	0.78	5.99
extra over for						
straight coupling	nr	0.18	2.79	0.55	0.50	3.84
reduced coupling, 10 × 8mm	nr	0.18	2.79	3.42	0.93	7.14
reducer, 10 × 8mm	nr	0.18	2.79	2.39	0.78	5.96
elbow, 10mm	nr	0.18	2.79	0.95	0.56	4.30
street elbow, 10mm	nr	0.18	2.79	2.78	0.84	6.41
equal tee, 10mm	nr	0.22	3.41	2.17	0.84	6.42
elbow, 10mm	nr	0.18	2.79	1.82	0.69	5.30
made bend	nr	0.18	2.79	0.00	0.42	3.21

	Unit	Labour hours	Hours cost £	Materials £	O & P £	Total £
15mm diameter, to timber	m	0.20	3.10	1.82	0.74	5.66
15mm diameter, plugged and screwed	m	0.22	3.41	2.10	0.83	6.34
extra over for						
reduced coupling, 15 × 8mm	nr	0.18	2.79	2.42	0.78	5.99
reduced coupling, 15 × 10mm	nr	0.18	2.79	2.36	0.77	5.92
reduced coupling, 15 × 12mm	nr	0.18	2.79	1.79	0.69	5.27
adaptor coupling, 15mm × 1/2in	nr	0.18	2.79	2.77	0.83	6.39
straight female connector, 15mm × 1/2in	nr	0.18	2.79	2.91	0.86	6.56
female reducing connector, 15mm × 3/8in	nr	0.18	2.79	5.79	1.29	9.87
female reducing connector, 15mm × 1/4in	nr	0.18	2.79	5.16	1.19	9.14
straight male connector, 15mm × 1/2in	nr	0.18	2.79	2.49	0.79	6.07
male reducing connector, 15mm × 3/4in	nr	0.18	2.79	3.38	0.93	7.10
tank connector, 15mm × 1/2in	nr	0.18	2.79	6.18	1.35	10.32
reducer, 15 × 8mm	nr	0.18	2.79	1.99	0.72	5.50
reducer, 15 × 10mm	nr	0.18	2.79	0.78	0.54	4.11
reducer, 15 × 12mm	nr	0.18	2.79	1.56	0.65	5.00
female adaptor, 15 × 1/2in	nr	0.18	2.79	5.00	1.17	8.96
male adaptor, 15 × 1/2in	nr	0.18	2.79	5.11	1.19	9.09
adaptor, 15 × 1/2in	nr	0.18	2.79	2.42	0.78	5.99
male elbow, 15mm × 1/2in	nr	0.18	2.79	4.95	1.16	8.90
reducing male elbow, 15mm × 3/8in	nr	0.18	2.79	6.97	1.46	11.22
female elbow, 15mm × 1/2in	nr	0.18	2.79	4.28	1.06	8.13
reducing female elbow, 15mm × 1/4in	nr	0.18	2.79	6.82	1.44	11.05
backplate elbow, 15mm × 1/2in	nr	0.18	2.79	11.01	2.07	15.87
flanged bend, 15mm × 1/2in	nr	0.18	2.79	5.18	1.20	9.17
flanged bend, 15mm × 3/4in	nr	0.18	2.79	11.65	2.17	16.61
slow bend	nr	0.18	2.79	0.83	0.54	4.16
return bend	nr	0.18	2.79	7.77	1.58	12.14
obtuse elbow	nr	0.18	2.79	0.99	0.57	4.35

	Unit	Labour hours	Hours cost £	Materials £	O & P £	Total £
Extra over 15mm diameter pipework (cont'd)						
tee, reduced branch largest end 15mm	nr	0.22	3.41	9.38	1.92	14.71
tee, both ends reduced, largest end 15mm	nr	0.22	3.41	8.75	1.82	13.98
sweep tee, 90 degrees, 15mm	nr	0.22	3.41	8.77	1.83	14.01
offset tee, 15mm	nr	0.22	3.41	11.59	2.25	17.25
double sweep tee, 15mm	nr	0.22	3.41	9.91	2.00	15.32
cross equal tee, 15mm	nr	0.22	3.41	13.12	2.48	19.01
stop end, 15mm	nr	0.18	2.79	3.39	0.93	7.11
tap connector, 15mm	nr	0.18	2.79	3.86	1.00	7.65
bent union adaptor, 15mm × 3/4in	nr	0.18	2.79	8.59	1.71	13.09
bent male union connector, 15mm × 1/2in	nr	0.18	2.79	10.11	1.94	14.84
bent female union connector, 15mm × 1/2in	nr	0.18	2.79	10.11	1.94	14.84
made bend	nr	0.18	2.79	0.00	0.42	3.21
22mm diameter, to timber	m	0.22	3.41	2.86	0.94	7.21
22mm diameter, plugged and screwed	m	0.24	3.72	3.11	1.02	7.85
extra over for						
straight coupling	nr	0.20	3.10	3.43	0.98	7.51
reduced coupling, 22 × 8mm	nr	0.20	3.10	3.43	0.98	7.51
reduced coupling, 22 × 10mm	nr	0.20	3.10	3.43	0.98	7.51
reduced coupling, 22 × 15mm	nr	0.20	3.10	2.31	0.81	6.22
straight female connector, 22mm × 3/4in	nr	0.20	3.10	4.21	1.10	8.41
female reducing connector, 22mm × 1/2in	nr	0.20	3.10	6.69	1.47	11.26
straight male connector, 22mm × 3/4in	nr	0.20	3.10	4.40	1.13	8.63
male reducing connector, 22mm × 1/2in	nr	0.20	3.10	6.93	1.50	11.53

	Unit	Labour hours	Hours cost £	Materials £	O & P £	Total £
Extra over 22mm diameter pipework (cont'd)						
tank connector, 22mm × 1in	nr	0.20	3.10	9.42	1.88	14.40
reducer, 22 × 12mm	nr	0.20	3.10	3.56	1.00	7.66
female adaptor, 22 × 3/4in	nr	0.20	3.10	9.42	1.88	14.40
male adaptor, 22 × 3/4in	nr	0.20	3.10	6.52	1.44	11.06
adaptor, 22 × 3/4in	nr	0.20	3.10	2.68	0.87	6.65
street elbow, 22mm	nr	0.20	3.10	2.11	0.78	5.99
male elbow, 22mm × 3/4in	nr	0.20	3.10	7.02	1.52	11.64
female elbow, 22mm × 3/4in	nr	0.20	3.10	6.18	1.39	10.67
reducing female elbow, 22mm × 1/2in	nr	0.20	3.10	7.14	1.54	11.78
backplate elbow, 22mm × 3/4in	nr	0.20	3.10	11.14	2.14	16.38
flanged bend, 22mm × 3/4in	nr	0.20	3.10	12.82	2.39	18.31
overflow bend, 22mm × 3/4in	nr	0.20	3.10	15.71	2.82	21.63
slow bend	nr	0.20	3.10	1.63	0.71	5.44
return bend	nr	0.20	3.10	15.29	2.76	21.15
obtuse elbow	nr	0.20	3.10	2.07	0.78	5.95
tee, reduced branch largest end 22mm	nr	0.24	3.72	6.83	1.58	12.13
tee, both ends reduced, largest end 22mm	nr	0.24	3.72	2.95	1.00	7.67
corner tee, 22mm	nr	0.24	3.72	14.24	2.69	20.65
sweep tee, 90 degrees, 22mm	nr	0.24	3.72	11.29	2.25	17.26
offset tee, 22mm	nr	0.24	3.72	13.96	2.65	20.33
double sweep tee, 22mm	nr	0.24	3.72	13.50	2.58	19.80
cross equal tee, 22mm	nr	0.24	3.72	14.66	2.76	21.14
stop end, 22mm	nr	0.20	3.10	2.38	0.82	6.30
tap connector, 22mm × 1/2in	nr	0.20	3.10	9.91	1.95	14.96
bent tap connector, 22mm × 1/2in	nr	0.20	3.10	12.09	2.28	17.47
bent tap connector, 22mm × 3/4in	nr	0.20	3.10	6.04	1.37	10.51
bent union adaptor, 22mm × 1in	nr	0.20	3.10	8.06	1.67	12.83
bent male union connector, 22mm × 3/4in	nr	0.20	3.10	9.92	1.95	14.97

	Unit	Labour hours	Hours cost £	Materials £	O & P £	Total £
Extra over 22mm diameter pipework (cont'd)						
bent female union connector, 22mm × 3/4in	nr	0.20	3.10	13.16	2.44	18.70
made bend	nr	0.20	3.10	0.00	0.47	3.57
28mm diameter, to timber	m	0.23	3.57	4.06	1.14	8.77
28mm diameter, plugged and screwed	m	0.25	3.88	4.33	1.23	9.44
extra over for						
straight coupling	nr	0.24	3.72	1.18	0.74	5.64
reduced coupling, 28 × 15mm	nr	0.24	3.72	5.28	1.35	10.35
reduced coupling, 28 × 22mm	nr	0.24	3.72	3.23	1.04	7.99
adaptor coupling, 28mm × 1in	nr	0.24	3.72	4.79	1.28	9.79
straight female connector, 28mm × 1in	nr	0.24	3.72	7.93	1.75	13.40
straight male connector, 28mm × 1in	nr	0.24	3.72	7.11	1.62	12.45
male reducing connector, 28mm × 3/4in	nr	0.24	3.72	10.91	2.19	16.82
tank connector, 28mm × 1in	nr	0.24	3.72	12.39	2.42	18.53
tank connector, 28mm × 1in	nr	0.24	3.72	12.39	2.42	18.53
reducer, 28 × 15mm	nr	0.24	3.72	2.61	0.95	7.28
reducer, 28 × 22mm	nr	0.24	3.72	1.99	0.86	6.57
female adaptor, 28mm × 1in	nr	0.24	3.72	10.75	2.17	16.64
male adaptor, 28mm × 1in	nr	0.24	3.72	10.89	2.19	16.80
adaptor, 28mm × 1in	nr	0.24	3.72	3.80	1.13	8.65
reducing elbow, 28 × 22mm	nr	0.24	3.72	9.22	1.94	14.88
street elbow, 28mm	nr	0.24	3.72	6.18	1.49	11.39
male elbow, 28mm × 1in	nr	0.24	3.72	10.57	2.14	16.43
female elbow, 28mm × 1in	nr	0.24	3.72	10.71	2.16	16.59
slow bend	nr	0.24	3.72	10.71	2.16	16.59
return bend	nr	0.24	3.72	19.53	3.49	26.74
obtuse elbow	nr	0.24	3.72	3.99	1.16	8.87
equal tee, 28mm	nr	0.24	3.72	5.25	1.35	10.32

	Unit	Labour hours	Hours cost £	Materials £	O & P £	Total £
Extra over 28mm diameter pipework (cont'd)						
tee, reduced branch largest end 28mm	nr	0.28	4.34	7.76	1.82	13.92
tee, one end and one branch reduced, largest end 28mm	nr	0.28	4.34	9.70	2.11	16.15
tee, both ends reduced, largest end 28mm	nr	0.28	4.34	20.03	3.66	28.03
sweep tee, 90 degrees, 28mm	nr	0.28	4.34	19.00	3.50	26.84
double sweep tee, 28mm	nr	0.28	4.34	20.52	3.73	28.59
cross equal tee, 28mm	nr	0.28	4.34	21.03	3.81	29.18
stop end, 28mm	nr	0.24	3.72	4.25	1.20	9.17
bent male union connector, 28mm × 1in	nr	0.24	3.72	18.81	3.38	25.91
bent female union connector, 28mm × 1in	nr	0.24	3.72	18.81	3.38	25.91
made bend	nr	0.24	3.72	0.00	0.56	4.28
35mm diameter, to timber	m	0.28	4.34	4.51	1.33	10.18
35mm diameter, plugged and screwed	m	0.30	4.65	4.79	1.42	10.86
extra over for						
straight coupling	nr	0.26	4.03	3.84	1.18	9.05
reduced coupling, 35 × 22mm	nr	0.26	4.03	7.46	1.72	13.21
reduced coupling, 35 × 28mm	nr	0.26	4.03	7.61	1.75	13.39
adaptor coupling, 35mm × 1 1/4in	nr	0.26	4.03	8.01	1.81	13.85
flanged connector	nr	0.26	4.03	58.42	9.37	71.82
straight female connector, 35mm × 1 1/4in	nr	0.26	4.03	13.42	2.62	20.07
straight male connector, 35mm × 1 1/4in	nr	0.26	4.03	15.87	2.99	22.89
tank connector, 35mm × 1 1/4in	nr	0.26	4.03	15.87	2.99	22.89
reducer, 35 × 15mm	nr	0.26	4.03	7.78	1.77	13.58
reducer, 35 × 22mm	nr	0.26	4.03	7.40	1.71	13.14
reducer, 35 × 28mm	nr	0.26	4.03	5.48	1.43	10.94

	Unit	Labour hours	Hours cost £	Materials £	O & P £	Total £
Extra over 35mm diameter pipework (cont'd)						
female adaptor, 35mm × 1 1/4in	nr	0.26	4.03	19.45	3.52	27.00
male adaptor, 35mm × 1 1/4in	nr	0.26	4.03	15.88	2.99	22.90
adaptor, 35mm × 1 1/4in	nr	0.26	4.03	4.45	1.27	9.75
elbow, 35mm	nr	0.26	4.03	8.23	1.84	14.10
street elbow, 35mm	nr	0.26	4.03	13.85	2.68	20.56
male elbow, 35mm × 1 1/4in	nr	0.26	4.03	14.94	2.85	21.82
female elbow, 35mm × 1 1/4in	nr	0.26	4.03	16.57	3.09	23.69
equal tee, 35mm	nr	0.30	4.65	13.40	2.71	20.76
tee, reduced branch largest end 35mm	nr	0.30	4.65	17.62	3.34	25.61
tee, one end and one branch reduced, largest end 35mm	nr	0.30	4.65	22.19	4.03	30.87
tee, both ends reduced, largest end 35mm	nr	0.30	4.65	28.02	4.90	37.57
sweep tee, 90 degrees, 35mm	nr	0.30	4.65	26.94	4.74	36.33
stop end, 35mm	nr	0.26	4.03	9.40	2.01	15.44
bent male union adaptor, 35mm × 1 1/4in	nr	0.26	4.03	30.67	5.21	39.91
bent female union adaptor, 35mm × 1 1/4in	nr	0.26	4.03	30.67	5.21	39.91
made bend	nr	0.26	4.03	0.00	0.60	4.63
42mm diameter, to timber	m	0.30	4.65	9.92	2.19	16.76
42mm diameter, plugged and screwed	m	0.32	4.96	10.14	2.27	17.37
extra over for						
straight coupling	nr	0.28	4.34	6.40	1.61	12.35
reduced coupling, 42 × 28mm	nr	0.28	4.34	12.74	2.56	19.64
reduced coupling, 42 × 35mm	nr	0.28	4.34	11.19	2.33	17.86
adaptor coupling, 42mm × 1 1/2in	nr	0.28	4.34	10.17	2.18	16.69

	Unit	Labour hours	Hours cost £	Materials £	O & P £	Total £
Extra over 42mm diameter pipework (cont'd)						
flanged connector	nr	0.28	4.34	69.82	11.12	85.28
straight female connector, 42mm × 1 1/2in	nr	0.28	4.34	17.84	3.33	25.51
straight male connector, 42mm × 1 1/2in	nr	0.28	4.34	16.13	3.07	23.54
tank connector, 42mm × 1 1/2in	nr	0.28	4.34	20.82	3.77	28.93
reducer, 42 × 15mm	nr	0.28	4.34	11.96	2.45	18.75
reducer, 42 × 22mm	nr	0.28	4.34	13.35	2.65	20.34
reducer, 42 × 28mm	nr	0.28	4.34	12.58	2.54	19.46
reducer, 42 × 35mm	nr	0.28	4.34	10.33	2.20	16.87
female adaptor, 42mm × 1 1/2in	nr	0.28	4.34	24.50	4.33	33.17
male adaptor, 42mm × 1 1/2in	nr	0.28	4.34	21.95	3.94	30.23
adaptor, 42mm × 1 1/2in	nr	0.28	4.34	5.86	1.53	11.73
elbow, 42mm	nr	0.28	4.34	13.61	2.69	20.64
street elbow, 42mm	nr	0.28	4.34	19.51	3.58	27.43
male elbow, 42mm × 1 1/2in	nr	0.28	4.34	17.77	3.32	25.43
equal tee, 42mm	nr	0.32	4.96	21.50	3.97	30.43
tee, reduced branch largest end 42mm	nr	0.32	4.96	40.57	6.83	52.36
tee, one end and one branch reduced, largest end 42mm	nr	0.32	4.96	37.95	6.44	49.35
tee, both ends reduced, largest end 42mm	nr	0.32	4.96	28.02	4.95	37.93
sweep tee, 90 degrees, 42mm	nr	0.32	4.96	39.94	6.74	51.64
stop end, 42mm	nr	0.28	4.34	16.18	3.08	23.60
bent male union adaptor, 42mm × 1 1/2in	nr	0.28	4.34	49.90	8.14	62.38
bent female union adaptor, 42mm × 1 1/2in	nr	0.28	4.34	49.90	8.14	62.38
made bend	nr	0.28	4.34	0.00	0.65	4.99

	Unit	Labour hours	Hours cost £	Materials £	O & P £	Total £
54mm diameter, to timber	m	0.32	4.96	14.01	2.85	21.82
54mm diameter, plugged and screwed	m	0.34	5.27	14.45	2.96	22.68
extra over for						
straight coupling	nr	0.30	4.65	11.82	2.47	18.94
reduced coupling, 54 × 35mm	nr	0.30	4.65	19.61	3.64	27.90
reduced coupling, 54 × 42mm	nr	0.30	4.65	21.38	3.90	29.93
straight female connector, 54mm × 2in	nr	0.30	4.65	28.30	4.94	37.89
straight male connector, 54mm × 2in	nr	0.30	4.65	24.51	4.37	33.53
tank connector, 54mm × 2in	nr	0.30	4.65	31.81	5.47	41.93
reducer, 54 × 15mm	nr	0.30	4.65	26.40	4.66	35.71
reducer, 54 × 22mm	nr	0.30	4.65	25.87	4.58	35.10
reducer, 54 × 28mm	nr	0.30	4.65	22.74	4.11	31.50
reducer, 54 × 35mm	nr	0.30	4.65	21.68	3.95	30.28
reducer, 54 × 42mm	nr	0.30	4.65	18.71	3.50	26.86
female adaptor, 54mm × 2in	nr	0.30	4.65	29.49	5.12	39.26
male adaptor, 54mm × 2in	nr	0.30	4.65	29.49	5.12	39.26
adaptor, 54mm × 2in	nr	0.30	4.65	12.84	2.62	20.11
elbow, 54mm	nr	0.30	4.65	13.84	2.77	21.26
male elbow, 54mm × 2in	nr	0.30	4.65	28.08	4.91	37.64
equal tee, 54mm	nr	0.34	5.27	31.43	5.51	42.21
tee, reduced branch largest end 54mm	nr	0.34	5.27	43.34	7.29	55.90
tee, one end and one branch reduced, largest end 54mm	nr	0.34	5.27	65.13	10.56	80.96
sweep tee, 90 degrees, 54mm	nr	0.34	5.27	57.54	9.42	72.23
stop end, 54mm	nr	0.34	5.27	41.36	6.99	53.62
bent union adaptor, 54mm × 54mm × 2in	nr	0.34	5.27	22.59	4.18	32.04
bent male union adaptor, 54mm × 2in	nr	0.34	5.27	78.81	12.61	96.69
bent female union adaptor, 54mm × 2in	nr	0.34	5.27	78.81	12.61	96.69

	Unit	Labour hours	Hours cost £	Materials £	O & P £	Total £
Stop valves						
Gunmetal stop valve with brass headwork, copper × copper						
15mm	nr	0.20	3.10	5.38	1.27	9.75
22mm	nr	0.22	3.41	10.04	2.02	15.47
28mm	nr	0.28	4.34	28.53	4.93	37.80
Dezincification-resistant stop valve, copper × copper						
15mm	nr	0.20	3.10	13.50	2.49	19.09
22mm	nr	0.22	3.41	23.41	4.02	30.84
28mm	nr	0.28	4.34	39.00	6.50	49.84
Gunmetal stop valve, copper × copper						
35mm	nr	0.32	4.96	61.17	9.92	76.05
42mm	nr	0.35	5.43	81.23	13.00	99.65
54mm	nr	0.42	6.51	121.34	19.18	147.03
Gunmetal lockshield stop valve with brass headwork, copper × copper						
22mm	nr	0.22	3.41	15.51	2.84	21.76
28mm	nr	0.28	4.34	28.45	4.92	37.71
Gunmetal union stop valve with brass headwork, copper × copper union						
15mm	nr	0.20	3.10	15.18	2.74	21.02
22mm	nr	0.22	3.41	22.29	3.86	29.56
Gunmetal double union lockshield stop valve with brass headwork, copper × copper union						
15mm	nr	0.20	3.10	17.84	3.14	24.08
22mm	nr	0.22	3.41	24.51	4.19	32.11
28mm	nr	0.28	4.34	43.60	7.19	55.13

	Unit	Labour hours	Hours cost £	Materials £	O & P £	Total £
Stop valves (cont'd)						
Dezincification-resistant double union lockshield stop valve, copper × copper						
15mm	nr	0.20	3.10	28.72	4.77	36.59
22mm	nr	0.22	3.41	34.96	5.76	44.13
28mm	nr	0.28	4.34	64.66	10.35	79.35
Dezincification-resistant double union stop valve with easy-clean cover, copper × copper						
15mm	nr	0.20	3.10	30.37	5.02	38.49
22mm	nr	0.22	3.41	37.91	6.20	47.52
28mm	nr	0.28	4.34	70.50	11.23	86.07
Gunmetal double union stop valve with easy-clean cover, copper × copper						
35mm	nr	0.35	5.43	107.59	16.95	129.97
42mm	nr	0.42	6.51	147.83	23.15	177.49
54mm	nr	0.46	7.13	232.55	35.95	275.63
Combined gunmetal stop valve and draincock with brass headwork, copper × copper						
15mm	nr	0.20	3.10	29.94	4.96	38.00
22mm	nr	0.22	3.41	36.79	6.03	46.23
Gate valves						
Gunmetal fullway gate valve, copper × copper						
15mm	nr	0.20	3.10	16.84	2.99	22.93
22mm	nr	0.22	3.41	19.51	3.44	26.36
28mm	nr	0.28	4.34	27.16	4.73	36.23
35mm	nr	0.32	4.96	60.61	9.84	75.41
42mm	nr	0.35	5.43	75.78	12.18	93.39
54mm	nr	0.42	6.51	109.93	17.47	133.91

	Unit	Labour hours	Hours cost £	Materials £	O & P £	Total £
Plug cocks						
Nited						
15mm	nr	0.20	3.10	12.49	2.34	17.93
22mm	nr	0.20	3.10	18.81	3.29	25.20
Nited with drop head						
15mm	nr	0.25	3.88	14.19	2.71	20.77
Nited with union outlet						
15mm	nr	0.25	3.88	14.49	2.75	21.12
22mm	nr	0.25	3.88	22.00	3.88	29.76
Nited union stopcock with drop head						
15mm	nr	0.25	3.88	16.43	3.05	23.35
22mm	nr	0.25	3.88	23.68	4.13	31.69
Valves						
Radiator valve, plain finish, compression inlet × taper male union outlet						
8mm × 1/2in	nr	0.30	4.65	18.12	3.42	26.19
10mm × 1/2in	nr	0.30	4.65	19.73	3.66	28.04
Radiator valve, chromium plated, compression inlet × taper male union outlet						
8mm × 1/2in	nr	0.30	4.65	22.02	4.00	30.67
10mm × 1/2in	nr	0.30	4.65	24.20	4.33	33.18
Radiator valve, plain finish, compression inlet × taper male union outlet						
8mm × 1/2in	nr	0.30	4.65	7.06	1.76	13.47
10mm × 1/2in	nr	0.30	4.65	7.06	1.76	13.47

	Unit	Labour hours	Hours cost £	Materials £	O & P £	Total £
Radiator valve, nickel plated, compression inlet × taper male union outlet						
8mm × 1/2in	nr	0.30	4.65	7.29	1.79	13.73
10mm × 1/2in	nr	0.30	4.65	7.29	1.79	13.73
Copper pipe, dezincification-resistant compression joints and fittings, clips at 1250mm maximum centres						
15mm diameter, to timber	m	0.26	4.03	1.84	0.88	6.75
15mm diameter, plugged and screwed	m	0.28	4.34	2.09	0.96	7.39
extra over for						
male coupling, 15mm × 3/4in	nr	0.18	2.79	5.22	1.20	9.21
tank coupling with backnut, 15mm × 1/2in	nr	0.18	2.79	6.09	1.33	10.21
female coupling, 15mm × 3/4in	nr	0.18	2.79	4.69	1.12	8.60
male elbow, 15mm × 1/2in	nr	0.18	2.79	3.95	1.01	7.75
male elbow, BSP parallel thread, 15mm × 1/2in	nr	0.18	2.79	3.95	1.01	7.75
female elbow, 15mm × 1/2in	nr	0.18	2.79	4.22	1.05	8.06
female elbow, 15mm × 3/4in	nr	0.18	2.79	7.59	1.56	11.94
female wall elbow, 15mm × 1/2in	nr	0.18	2.79	6.16	1.34	10.29
tee with reduced branch, 15 × 15 × 12mm	nr	0.22	3.41	7.85	1.69	12.95
tee, end reduced 15 × 12 × 12mm	nr	0.22	3.41	5.60	1.35	10.36
tee, end and branch reduced, 15 × 12 × 12mm	nr	0.22	3.41	6.07	1.42	10.90
tee, both ends reduced, 15 × 15 × 18mm	nr	0.22	3.41	11.17	2.19	16.77
tee, both ends reduced, 15 × 15 × 22mm	nr	0.22	3.41	8.77	1.83	14.01
female tee, 15mm × 1/2in × 15mm	nr	0.22	3.41	8.00	1.71	13.12

	Unit	Labour hours	Hours cost £	Materials £	O & P £	Total £
Extra over 15mm diameter pipework (cont'd)						
male tee, 15mm × 1/2in × 15mm	nr	0.22	3.41	9.16	1.89	14.46
cross tee, 15mm	nr	0.22	3.41	10.77	2.13	16.31
straight swivel connector, 15mm × 3/4in	nr	0.18	2.79	6.94	1.46	11.19
bent swivel connector, 15mm × 1/2in	nr	0.18	2.79	5.15	1.19	9.13
18mm diameter, to timber	m	0.21	3.26	2.62	0.88	6.76
18mm diameter, plugged and screwed	m	0.23	3.57	2.89	0.97	7.42
extra over for						
straight coupling, 18mm	nr	0.18	2.79	5.31	1.22	9.32
male coupling, 18mm × 1/2in	nr	0.18	2.79	4.30	1.06	8.15
female coupling, 18mm × 1/2in	nr	0.18	2.79	4.69	1.12	8.60
elbow, 18mm	nr	0.18	2.79	5.93	1.31	10.03
male elbow, 18mm × 1/2in	nr	0.18	2.79	5.83	1.29	9.91
female elbow, 18mm × 1/2in	nr	0.18	2.79	6.66	1.42	10.87
tee, 18mm	nr	0.22	3.41	8.37	1.77	13.55
tee with reduced branch, 18 × 15 × 12mm	nr	0.22	3.41	7.85	1.69	12.95
tee with reduced branch, 18 × 18 × 15mm	nr	0.22	3.41	10.06	2.02	15.49
tee, end reduced 18 × 12 × 18mm	nr	0.22	3.41	10.99	2.16	16.56
tee, end reduced 18 × 15 × 18mm	nr	0.22	3.41	10.06	2.02	15.49
tee, end and branch reduced, 18 × 12 × 12mm	nr	0.22	3.41	10.06	2.02	15.49
tee, end and branch reduced, 18 × 15 × 12mm	nr	0.22	3.41	10.06	2.02	15.49
tee, end and branch reduced, 18 × 15 × 12mm	nr	0.22	3.41	9.87	1.99	15.27
reducing set, 18 × 12mm	nr	0.18	2.79	2.60	0.81	6.20
reducing set, 18 × 15mm	nr	0.18	2.79	2.60	0.81	6.20

	Unit	Labour hours	Hours cost £	Materials £	O & P £	Total £
22mm diameter, to timber	m	0.22	3.41	3.21	0.99	7.61
22mm diameter, plugged and screwed	m	0.24	3.72	3.56	1.09	8.37
extra over for						
male coupling, 22mm × 1/2in	nr	0.18	2.79	4.60	1.11	8.50
tank coupling with backnut, 22mm × 3/4in	nr	0.18	2.79	6.96	1.46	11.21
tank coupling with backnut, 22mm × 1in	nr	0.18	2.79	11.59	2.16	16.54
female coupling, 22mm × 1/2in	nr	0.18	2.79	5.01	1.17	8.97
female coupling, 22mm × 1in	nr	0.18	2.79	6.80	1.44	11.03
male elbow, 22mm × 3/4in	nr	0.18	2.79	4.71	1.13	8.63
male elbow, 22mm × 1in	nr	0.18	2.79	9.66	1.87	14.32
male elbow, BSP parallel thread, 22mm × 3/4in	nr	0.18	2.79	4.71	1.13	8.63
male elbow, BSP parallel thread, 22mm × 1in	nr	0.18	2.79	9.66	1.87	14.32
female elbow, 22mm × 3/4in	nr	0.18	2.79	6.09	1.33	10.21
female elbow, 22mm × 1in	nr	0.18	2.79	9.77	1.88	14.44
female wall elbow, 22mm × 3/4in	nr	0.18	2.79	11.62	2.16	16.57
tee with reduced branch, 22 × 22 × 12mm	nr	0.22	3.41	11.69	2.27	17.37
tee with reduced branch, 22 × 22 × 15mm	nr	0.22	3.41	8.45	1.78	13.64
tee, end reduced 22 × 15 × 22mm	nr	0.22	3.41	9.82	1.98	15.21
tee, end and branch reduced, 22 × 15 × 15mm	nr	0.22	3.41	9.24	1.90	14.55
tee, both ends reduced, 22 × 22 × 28mm	nr	0.22	3.41	14.72	2.72	20.85
female tee, 22mm × 1/2in × 22mm	nr	0.22	3.41	10.07	2.02	15.50
female tee, 22mm × 3/4in × 22mm	nr	0.22	3.41	10.27	2.05	15.73
male end tee, 22mm × 3/4in × 22mm	nr	0.22	3.41	12.95	2.45	18.81

	Unit	Labour hours	Hours cost £	Materials £	O & P £	Total £
Extra over 22mm diameter pipework (cont'd)						
straight swivel connector, 22mm × 3/4in	nr	0.18	2.79	7.59	1.56	11.94
bent swivel connector, 22mm × 3/4in	nr	0.18	2.79	9.15	1.79	13.73
male adaptor, 22mm × 1/2in	nr	0.18	2.79	5.90	1.30	9.99
male adaptor, 22mm × 3/4in	nr	0.18	2.79	5.90	1.30	9.99
female adaptor, 22mm × 1/2in	nr	0.18	2.79	5.90	1.30	9.99
female adaptor, 22mm × 3/4in	nr	0.18	2.79	4.28	1.06	8.13
reducing set, 22 × 12mm	nr	0.18	2.79	3.18	0.90	6.87
reducing set, 22 × 15mm	nr	0.18	2.79	2.38	0.78	5.95
reducing set, 22 × 18mm	nr	0.18	2.79	2.73	0.83	6.35
28mm diameter, to timber	m	0.23	3.57	3.7	1.09	8.35
28mm diameter, plugged and screwed	m	0.25	3.88	3.95	1.17	9.00
extra over for						
straight coupling, 28mm	nr	0.20	3.10	6.99	1.51	11.60
male coupling, 28mm × 3/4in	nr	0.20	3.10	7.28	1.56	11.94
tank coupling with backnut, 28mm × 1in	nr	0.20	3.10	11.19	2.14	16.43
female coupling, 28mm × 1in	nr	0.20	3.10	6.43	1.43	10.96
elbow, 28mm	nr	0.20	3.10	9.05	1.82	13.97
male elbow, 28mm × 1in	nr	0.20	3.10	9.03	1.82	13.95
male elbow, BSP parallel thread, 28mm × 1in	nr	0.20	3.10	8.28	1.71	13.09
female elbow, 28mm × 1in	nr	0.20	3.10	10.10	1.98	15.18
tee, 28mm	nr	0.24	3.72	14.42	2.72	20.86
tee with reduced branch, 28 × 28 × 15mm	nr	0.24	3.72	13.93	2.65	20.30
tee with reduced branch, 28 × 28 × 22mm	nr	0.24	3.72	13.93	2.65	20.30
tee, end reduced 28 × 28 × 15mm	nr	0.24	3.72	16.11	2.97	22.80
tee, end reduced 28 × 28 × 22mm	nr	0.24	3.72	15.26	2.85	21.83

	Unit	Labour hours	Hours cost £	Materials £	O & P £	Total £
Extra over 28mm diameter pipework (cont'd)						
tee, both ends reduced, 28 × 28 × 22mm	nr	0.24	3.72	15.10	2.82	21.64
female tee, 28mm × 1/2in × 28mm	nr	0.24	3.72	16.79	3.08	23.59
female tee, 28mm × 1in × 28mm	nr	0.24	3.72	16.79	3.08	23.59
female adaptor, 28mm × 1in	nr	0.20	3.10	9.71	1.92	14.73
reducing set, 28 × 15mm	nr	0.20	3.10	3.56	1.00	7.66
reducing set, 28 × 22mm	nr	0.20	3.10	3.56	1.00	7.66
stop end, 28mm	nr	0.20	3.10	7.31	1.56	11.97
35mm diameter, to timber	m	0.28	4.34	7.77	1.82	13.93
35mm diameter, plugged and screwed	m	0.30	4.65	8.03	1.90	14.58
extra over for						
straight coupling, 35mm	nr	0.26	4.03	14.55	2.79	21.37
male coupling, 35mm × 1 in	nr	0.26	4.03	18.01	3.31	25.35
male coupling, 35mm × 1 1/4in	nr	0.26	4.03	11.05	2.26	17.34
tank coupling with backnut, 35mm × 1 1/4in	nr	0.26	4.03	13.31	2.60	19.94
female coupling, 35mm × 1 1/4in	nr	0.26	4.03	13.27	2.60	19.90
elbow, 35mm	nr	0.26	4.03	19.63	3.55	27.21
male elbow, 35mm × 1 1/4in	nr	0.26	4.03	18.26	3.34	25.63
female elbow, 35mm × 1 1/4in	nr	0.26	4.03	17.44	3.22	24.69
tee, 35mm	nr	0.30	4.65	25.56	4.53	34.74
tee with reduced branch, 35 × 35 × 15mm	nr	0.30	4.65	24.97	4.44	34.06
tee with reduced branch, 35 × 35 × 22mm	nr	0.30	4.65	24.97	4.44	34.06
tee with reduced branch, 35 × 35 × 28mm	nr	0.30	4.65	24.97	4.44	34.06
reducing set, 35 × 15mm	nr	0.26	4.03	6.10	1.52	11.65

	Unit	Labour hours	Hours cost £	Materials £	O & P £	Total £
Extra over 35mm diameter pipework (cont'd)						
reducing set, 35 × 22mm	nr	0.26	4.03	6.10	1.52	11.65
reducing set, 35 × 28mm	nr	0.26	4.03	6.10	1.52	11.65
tank connector, 35mm	nr	0.26	4.03	8.81	1.93	14.77
stop end, 35mm	nr	0.26	4.03	11.27	2.30	17.60
42mm diameter, to timber	m	0.30	4.65	9.77	2.16	16.58
42mm diameter, plugged and screwed	m	0.32	4.96	10.03	2.25	17.24
extra over for						
straight coupling, 42mm	nr	0.28	4.34	19.13	3.52	26.99
male coupling, 42mm × 1 1/2in	nr	0.28	4.34	16.21	3.08	23.63
tank coupling with backnut, 42mm × 1 1/2in	nr	0.28	4.34	17.90	3.34	25.58
female coupling, 42mm × 1 1/2in	nr	0.28	4.34	17.86	3.33	25.53
elbow, 42mm	nr	0.28	4.34	26.60	4.64	35.58
male elbow, 42mm × 1 1/2in	nr	0.28	4.34	26.80	4.67	35.81
female elbow, 42mm × 1 1/2in	nr	0.28	4.34	26.81	4.67	35.82
tee, 42mm	nr	0.32	4.96	40.16	6.77	51.89
tee with reduced branch, 42 × 42 × 22mm	nr	0.32	4.96	38.59	6.53	50.08
tee with reduced branch, 42 × 42 × 28mm	nr	0.32	4.96	38.59	6.53	50.08
reducing set, 42 × 15mm	nr	0.28	4.34	11.13	2.32	17.79
reducing set, 42 × 22mm	nr	0.28	4.34	9.65	2.10	16.09
reducing set, 42 × 28mm	nr	0.28	4.34	9.65	2.10	16.09
reducing set, 42mm × 35mm	nr	0.28	4.34	9.65	2.10	16.09
stop end, 42mm	nr	0.28	4.34	18.76	3.47	26.57

	Unit	Labour hours	Hours cost £	Materials £	O & P £	Total £
54mm diameter, to timber	m	0.32	4.96	12.15	2.57	19.68
54mm diameter, plugged and screwed	m	0.34	5.27	12.64	2.69	20.60
extra over for						
straight coupling, 54mm	nr	0.30	4.65	28.61	4.99	38.25
male coupling, 54mm × 2in	nr	0.30	4.65	26.20	4.63	35.48
tank coupling with backnut, 54mm × 2in	nr	0.30	4.65	24.50	4.37	33.52
female coupling, 54mm × 2in	nr	0.30	4.65	26.20	4.63	35.48
elbow, 54mm	nr	0.30	4.65	45.78	7.56	57.99
male elbow, 54mm × 2in	nr	0.30	4.65	41.47	6.92	53.04
tee, 54mm	nr	0.34	5.27	64.53	10.47	80.27
tee with reduced branch, 54 × 54 × 28mm	nr	0.34	5.27	64.53	10.47	80.27
reducing set, 54 × 22mm	nr	0.30	4.65	18.48	3.47	26.60
reducing set, 54 × 28mm	nr	0.30	4.65	18.48	3.47	26.60
reducing set, 54 × 35mm	nr	0.30	4.65	16.10	3.11	23.86
reducing set, 54 × 42mm	nr	0.30	4.65	16.10	3.11	23.86
Copper pipe, compression joints and fittings, clips at 1250mm centres						
15mm diameter, to timber	m	0.20	3.10	1.75	0.73	5.58
15mm diameter, plugged and screwed	m	0.22	3.41	2.05	0.82	6.28
extra over for						
bent union radiator, 15mm × 1/2in	nr	0.18	2.79	6.99	1.47	11.25
air-release elbow, 15mm	nr	0.18	2.79	9.03	1.77	13.59
slow bend, 15mm	nr	0.18	2.79	9.03	1.77	13.59
male elbow, 15mm × 1/2in	nr	0.18	2.79	2.45	0.79	6.03
male elbow, 15mm × 3/4in	nr	0.18	2.79	7.47	1.54	11.80
male elbow, BSP parallel thread, 15mm × 1/2in	nr	0.18	2.79	2.45	0.79	6.03
female elbow, 15mm × 1/2in	nr	0.18	2.79	3.76	0.98	7.53
offset tee, 15 × 15 × 15mm	nr	0.18	2.79	16.25	2.86	21.90

	Unit	Labour hours	Hours cost £	Materials £	O & P £	Total £
Extra over 15mm diameter pipework (cont'd)						
male adaptor, 15mm × 1/2in	nr	0.18	2.79	2.97	0.86	6.62
reducing set, 15 × 8mm	nr	0.18	2.79	2.02	0.72	5.53
reducing set, 15 × 10mm	nr	0.18	2.79	2.02	0.72	5.53
reducing set, 15 × 12mm	nr	0.18	2.79	2.02	0.72	5.53
22mm diameter, to timber	m	0.21	3.26	2.63	0.88	6.77
22mm diameter, plugged and screwed	m	0.23	3.57	2.82	0.96	7.34
extra over for						
lead coupling, 22mm	nr	0.18	2.79	3.89	1.00	7.68
air-release elbow, 22mm	nr	0.18	2.79	11.95	2.21	16.95
slow bend, 22mm	nr	0.18	2.79	14.20	2.55	19.54
male elbow, 22mm × 3/4in	nr	0.18	2.79	3.16	0.89	6.84
male elbow, 22mm × 1in	nr	0.18	2.79	6.19	1.35	10.33
male elbow, BSP parallel thread, 22mm × 3/4in	nr	0.18	2.79	3.16	0.89	6.84
male elbow, BSP parallel thread, 22mm × 1in	nr	0.18	2.79	6.19	1.35	10.33
female elbow, 22mm × 3/4in	nr	0.18	2.79	5.44	1.23	9.46
female elbow, 22mm × 1in	nr	0.18	2.79	8.47	1.69	12.95
bent cylinder connector, 22mm × 1in	nr	0.18	2.79	12.45	2.29	17.53
tee with reduced branch, 22 × 22 × 15mm	nr	0.22	3.41	7.84	1.69	12.94
tee, end reduced 22 × 15 × 22mm	nr	0.22	3.41	9.12	1.88	14.41
tee, end and branch reduced, 22 × 15 × 15mm	nr	0.22	3.41	9.12	1.88	14.41
straight swivel connector, 22mm × 1/2in	nr	0.18	2.79	7.35	1.52	11.66
bent swivel connector, 22mm × 1/2in	nr	0.18	2.79	9.71	1.88	14.38
offset tee, 22 × 22 × 15mm	nr	0.22	3.41	23.91	4.10	31.42

	Unit	Labour hours	Hours cost £	Materials £	O & P £	Total £
28mm diameter, to timber	m	0.22	3.41	3.50	1.04	7.95
28mm diameter, plugged and screwed	m	0.24	3.72	3.72	1.12	8.56
extra over for						
straight coupling, 28mm	nr	0.20	3.10	6.02	1.37	10.49
male coupling, 28mm × 1 1/4in	nr	0.20	3.10	13.56	2.50	19.16
female coupling, 28mm × 3/4in	nr	0.20	3.10	7.90	1.65	12.65
female coupling, 28mm × 1in	nr	0.20	3.10	4.77	1.18	9.05
slow bend, 28mm	nr	0.20	3.10	19.13	3.33	25.56
male elbow, 28mm × 1in	nr	0.20	3.10	6.19	1.39	10.68
male elbow, BSP parallel thread, 28mm × 1in	nr	0.20	3.10	7.69	1.62	12.41
female elbow, 28mm × 1in	nr	0.20	3.10	9.57	1.90	14.57
tee, 28 × 28 × 28mm	nr	0.24	3.72	13.85	2.64	20.21
Stop valves						
Gunmetal stop valve with brass headwork, copper × copper						
15mm	nr	0.19	2.95	7.38	1.55	11.87
22mm	nr	0.21	3.26	12.97	2.43	18.66
28mm	nr	0.27	4.19	33.84	5.70	43.73
Gunmetal lockshield stop valve with brass headwork, copper × copper						
15mm	nr	0.19	2.95	16.05	2.85	21.84
22mm	nr	0.21	3.26	22.65	3.89	29.79
Dezincification-resistant stop valve, copper × copper						
15mm	nr	0.19	2.95	17.20	3.02	23.17
22mm	nr	0.21	3.26	28.24	4.72	36.22
28mm	nr	0.27	4.19	46.74	7.64	58.56

	Unit	Labour hours	Hours cost £	Materials £	O & P £	Total £
Dezincification-resistant lockshield stop valve, copper × copper						
15mm	nr	0.19	2.95	19.78	3.41	26.13
22mm	nr	0.21	3.26	30.97	5.13	39.36
Gunmetal stop valve with brass headwork, polyethylene × copper						
20 × 15mm	nr	0.19	2.95	17.52	3.07	23.53
25 × 15mm	nr	0.21	3.26	29.89	4.97	38.12
25 × 22mm	nr	0.27	4.19	33.71	5.68	43.58
Gate valves						
Brass gate valve, copper × copper						
15mm	nr	0.19	2.95	12.04	2.25	17.23
22mm	nr	0.21	3.26	14.04	2.59	19.89
28mm	nr	0.27	4.19	24.52	4.31	33.01
Brass lockshield gate valve, copper × copper						
15mm	nr	0.19	2.95	12.15	2.26	17.36
22mm	nr	0.21	3.26	14.32	2.64	20.21
28mm	nr	0.27	4.19	25.96	4.52	34.67
Gunmetal fullway gate valve, copper × copper						
15mm	nr	0.19	2.95	21.93	3.73	28.61
22mm	nr	0.21	3.26	24.67	4.19	32.11
28mm	nr	0.27	4.19	33.53	5.66	43.37
35mm	nr	0.36	5.58	48.64	8.13	62.35
42mm	nr	0.50	7.75	67.38	11.27	86.40
54mm	nr	0.60	9.30	92.75	15.31	117.36

	Unit	Labour hours	Hours cost £	Materials £	O & P £	Total £
Gunmetal lockshield gate valve, copper × copper						
15mm	nr	0.19	2.95	21.51	3.67	28.12
22mm	nr	0.21	3.26	25.51	4.31	33.08
28mm	nr	0.27	4.19	38.08	6.34	48.60
35mm	nr	0.36	5.58	51.32	8.54	65.44
42mm	nr	0.50	7.75	70.68	11.76	90.19
54mm	nr	0.60	9.30	97.49	16.02	122.81
Check valves						
Single check valve with female BSP parallel threads						
1/2in	nr	0.19	2.95	15.05	2.70	20.69
3/4in	nr	0.21	3.26	24.01	4.09	31.35
1in	nr	0.27	4.19	32.77	5.54	42.50
Double check valve with female BSP parallel threads						
1/2in	nr	0.19	2.95	17.04	3.00	22.98
3/4in	nr	0.21	3.26	35.09	5.75	44.10
1in	nr	0.27	4.19	52.61	8.52	65.31
Drainage fittings						
Dezincification-resistant stop valve adaptor, copper × copper						
15mm	nr	0.20	3.10	11.72	2.22	17.04
Draw-off coupling, copper × copper						
15mm	nr	0.20	3.10	10.71	2.07	15.88
Dezincification-resistant draw-off coupling, copper × copper						
15mm	nr	0.20	3.10	14.78	2.68	20.56
22mm	nr	0.25	3.88	17.10	3.15	24.12

	Unit	Labour hours	Hours cost £	Materials £	O & P £	Total £
Dezincification-resistant draw-off elbow, copper × parallel main thread						
15mm × 1/2in	nr	0.20	3.10	11.28	2.16	16.54
22mm × 3/4in	nr	0.25	3.88	12.66	2.48	19.02
28mm × 1in	nr	0.30	4.65	20.67	3.80	29.12
Brass draw-off elbow, copper × copper						
15mm	nr	0.20	3.10	10.95	2.11	16.16
22mm	nr	0.25	3.88	12.56	2.47	18.90
28mm	nr	0.30	4.65	19.35	3.60	27.60
Dezincification-resistant draw-off elbow, copper × copper						
15mm	nr	0.20	3.10	16.23	2.90	22.23
22mm	nr	0.25	3.88	19.44	3.50	26.81
Servicing valves						
Straight pattern, copper × copper						
15 × 15mm	nr	0.20	3.10	5.75	1.33	10.18
22 × 22mm	nr	0.25	3.88	12.77	2.50	19.14
Straight pattern, copper × swivel end						
15mm × 1/2in	nr	0.20	3.10	6.50	1.44	11.04
Appliance valves						
Chromium-plated stop valve						
15mm × 3/4in	nr	0.20	3.10	4.62	1.16	8.88
Brass tee valve						
15mm × 3/4in	nr	0.20	3.10	6.25	1.40	10.75

	Unit	Labour hours	Hours cost £	Materials £	O & P £	Total £
Cold water storage tanks						
Galvanised steel cistern with cover, reference						
SC10, 18 litres	nr	0.60	9.30	52.66	9.29	71.25
SC15, 36 litres	nr	0.65	10.08	66.36	11.47	87.90
SC20, 54 litres	nr	0.65	10.08	69.30	11.91	91.28
SC25, 68 litres	nr	0.70	10.85	73.03	12.58	96.46
SC30, 86 litres	nr	0.75	11.63	81.39	13.95	106.97
SC40, 114 litres	nr	0.80	12.40	87.94	15.05	115.39
SC50, 154 litres	nr	0.85	13.18	130.44	21.54	165.16
Cutting holes for connectors						
13mm	nr	0.10	1.55	0.00	0.23	1.78
18mm	nr	0.12	1.86	0.00	0.28	2.14
25mm	nr	0.15	2.33	0.00	0.35	2.67
32mm	nr	0.20	3.10	0.00	0.47	3.57
Plastic cistern with lid, reference						
PC4, 18 litres	nr	0.60	9.30	7.29	2.49	19.08
PC15, 68 litres	nr	0.70	10.85	17.59	4.27	32.71
PC20, 91 litres	nr	0.75	11.63	17.92	4.43	33.98
PC25, 114 litres	nr	0.80	12.40	23.83	5.43	41.66
PC40, 182 litres	nr	0.90	13.95	42.61	8.48	65.04
PC50, 227 litres	nr	1.00	15.50	42.93	8.76	67.19
Cutting holes for connectors						
13mm	nr	0.10	1.55	0.00	0.23	1.78
18mm	nr	0.12	1.86	0.00	0.28	2.14
25mm	nr	0.15	2.33	0.00	0.35	2.67
32mm	nr	0.20	3.10	0.00	0.47	3.57

	Unit	Labour hours	Hours cost £	Materials £	O & P £	Total £
Hot water copper cylinders						
Direct cylinders, insulated, reference						
ref 5, 98 litres	nr	0.50	7.75	143.63	22.71	174.09
ref 7, 120 litres	nr	0.55	8.53	151.86	24.06	184.44
ref 8, 148 litres	nr	0.65	10.08	167.00	26.56	203.64
ref 9, 166 litres	nr	0.85	13.18	182.29	29.32	224.78
Indirect cylinders, insulated, reference						
ref 2, 96 litres	nr	0.50	7.75	105.37	16.97	130.09
ref 3, 114 litres	nr	0.55	8.53	114.60	18.47	141.59
ref 7, 117 litres	nr	0.55	8.53	120.50	19.35	148.38
ref 8, 140 litres	nr	0.65	10.08	137.39	22.12	169.58
ref 9, 162 litres	nr	0.70	10.85	170.98	27.27	209.10
Insulation						
Preformed pipe lagging, fire-retardant foam 13mm thick for pipe size						
15mm	m	0.08	1.24	1.18	0.36	2.78
22mm	m	0.09	1.40	1.55	0.44	3.39
28mm	m	0.10	1.55	1.84	0.51	3.90
Preformed pipe lagging, fire-retardant foam 25mm thick for pipe size						
15mm	m	0.10	1.55	4.56	0.92	7.03
22mm	m	0.11	1.71	5.02	1.01	7.73
28mm	m	0.12	1.86	5.77	1.14	8.77
50mm glass-fibre filled insulating jacket, fixing bands to tank size						
445 × 305 × 300mm	nr	0.35	5.43	4.55	1.50	11.47
495 × 368 × 362mm	nr	0.40	6.20	5.57	1.77	13.54
630 × 450 × 420mm	nr	0.45	6.98	6.45	2.01	15.44
665 × 490 × 515mm	nr	0.50	7.75	7.56	2.30	17.61
995 × 605 × 595mm	nr	0.70	10.85	11.11	3.29	25.25

	Unit	Labour hours	Hours cost £	Materials £	O & P £	Total £
T10 GAS/OIL-FIRED BOILERS						
Gas-fired wall-mounted boilers for domestic central heating and indirect hot water						
Balanced flue, output						
30,000 BTU	nr	4.80	74.40	450.97	78.81	604.18
40,000 BTU	nr	4.80	74.40	519.71	89.12	683.23
50,000 BTU	nr	4.80	74.40	592.08	99.97	766.45
60,000 BTU	nr	4.80	74.40	686.25	114.10	874.75
Fan flue, output						
30,000 BTU	nr	4.20	65.10	481.20	81.95	628.25
40,000 BTU	nr	4.20	65.10	541.93	91.05	698.08
50,000 BTU	nr	4.20	65.10	591.31	98.46	754.87
60,000 BTU	nr	4.20	65.10	639.99	105.76	810.85
70,000 BTU	nr	4.20	65.10	744.21	121.40	930.71
80,000 BTU	nr	4.20	65.10	1143.81	181.34	1390.25
Gas-fired floor-standing boilers for domestic central heating and indirect hot water						
Fanned balanced flue, output						
40,000 BTU	nr	4.80	74.40	596.39	100.62	771.41
50,000 BTU	nr	4.80	74.40	611.60	102.90	788.90
60,000 BTU	nr	4.80	74.40	633.57	106.20	814.17
70,000 BTU	nr	4.80	74.40	666.89	111.19	852.48
80,000 BTU	nr	4.80	74.40	916.27	148.60	1139.27
Fanned chimney flue, output						
40,000 BTU	nr	4.20	65.10	466.17	79.69	610.96
50,000 BTU	nr	4.20	65.10	509.86	86.24	661.20
60,000 BTU	nr	4.20	65.10	561.70	94.02	720.82
70,000 BTU	nr	4.20	65.10	732.47	119.64	917.21
80,000 BTU	nr	4.20	65.10	934.07	149.88	1149.05

	Unit	Labour hours	Hours cost £	Materials £	O & P £	Total £
Gas-fired wall-mounted combination boiler for domestic central heating and hot water						
Fan flue, output						
30,400-81,900 BTU	nr	4.80	74.40	694.29	115.30	883.99
35,500-95,000 BTU	nr	4.80	74.40	745.34	122.96	942.70
Condensing fan flue						
25,200-92,800 BTU	nr	4.80	74.40	872.77	142.08	1089.25
25,900-93,800 BTU	nr	4.80	74.40	982.73	158.57	1215.70
Conventional flue						
31,000-82,000 BTU	nr	4.80	74.40	713.34	118.16	905.90
Gas-fired floor-standing combination boiler for domestic central heating and hot water						
Fan flue, output						
37,500-81,900 BTU	nr	4.80	74.40	1149.44	183.58	1407.42
Oil-fired boiler for domestic central heating and hot water						
Balanced flue, output						
40,000-48,000 BTU	nr	4.80	74.40	806.78	11.16	85.56
50,000-65,000 BTU	nr	4.80	74.40	902.49	204.52	1568.00
75,000-85,000 BTU	nr	4.80	74.40	1008.20	221.29	1696.58
88,000-110,000 BTU	nr	4.80	74.40	1130.59	243.78	1868.99
Conventional flue, output						
52,000 BTU	nr	4.80	74.40	806.78	132.18	1013.36
70,000 BTU	nr	4.80	74.40	902.81	146.58	1123.79
90,000 BTU	nr	4.80	74.40	1008.20	162.39	1244.99
120,000 BTU	nr	4.80	74.40	1130.59	180.75	1385.74

	Unit	Labour hours	Hours cost £	Materials £	O & P £	Total £
Oil-fired combination boiler for domestic central heating and hot water						
50,000-65,000 BTU	nr	4.80	74.40	1289.08	204.52	1568.00
75,000-85,000 BTU	nr	4.80	74.40	1400.89	221.29	1696.58
88,000-110,000 BTU	nr	4.80	74.40	1550.81	243.78	1868.99
Conventional flue, output						
52,000 BTU	nr	4.80	74.40	1218.20	193.89	1486.49
70,000 BTU	nr	4.80	74.40	1289.08	204.52	1568.00
90,000 BTU	nr	4.80	74.40	1400.89	221.29	1696.58
120,000 BTU	nr	4.80	74.40	1550.81	243.78	1868.99
Oil storage tanks						
Standard domestic pattern plastic oil storage tank						
1380 × 1780 × 890mm capacity 1200 litres	nr	1.50	23.25	182.41	30.85	236.51
745 × 1700 × 1380mm capacity 1225 litres	nr	1.50	23.25	258.86	42.32	324.43
1190mm diameter × 1400mm high, capacity 1300 litres	nr	1.50	23.25	182.82	30.91	236.98
1130 × 2055 × 1145mm capacity 1800 litres	nr	1.75	27.13	320.45	52.14	399.71
1265 × 2250 × 1320mm capacity 2500 litres	nr	2.00	31.00	399.16	64.52	494.68
1640mm diameter × 1750mm high, capacity 3600 litres	nr	2.50	38.75	440.21	71.84	550.80
2480mm diameter × 2640 high, capacity 10,000 litres	nr	3.00	46.50	1272.91	197.91	1517.32
Standard domestic pattern plastic bunded oil storage tank						
1380 × 1780 × 890mm capacity capacity1200 litres	nr	1.50	23.25	244.24	40.12	307.61
745 × 1700 × 1380mm capacity 1225 litres	nr	1.50	23.25	320.29	51.53	395.07

	Unit	Labour hours	Hours cost £	Materials £	O & P £	Total £
1190mm diameter × 1400 high, capacity 1300 litres	nr	1.50	23.25	748.73	115.80	887.78
1130 × 2055 × 1145mm capacity 1800 litres	nr	1.75	27.13	781.72	121.33	930.17
1265 × 2250 × 1320mm capacity 2500 litres	nr	2.00	31.00	868.15	134.87	1034.02

	Unit	Labour hours	Hours cost £	Materials £	O & P £	Total £
T30 RADIATORS						
Single convector radiator fixed to concealed brackets plugged and screwed to brickwork						
300mm high, length						
400mm	nr	0.90	13.95	16.00	4.49	34.44
800mm	nr	1.05	16.28	26.57	6.43	49.27
1200mm	nr	1.15	17.83	38.23	8.41	64.46
1600mm	nr	1.30	20.15	49.88	10.50	80.53
2000mm	nr	1.40	21.70	60.46	12.32	94.48
450mm high, length						
500mm	nr	1.00	15.50	15.65	4.67	35.82
1000mm	nr	1.15	17.83	27.33	6.77	51.93
1400mm	nr	1.25	19.38	36.47	8.38	64.22
1800mm	nr	1.40	21.70	52.65	11.15	85.50
2000mm	nr	1.50	23.25	70.15	14.01	107.41
600mm high, length						
500mm	nr	1.10	17.05	19.31	5.45	41.81
1000mm	nr	1.25	19.38	33.20	7.89	60.46
1400mm	nr	1.35	20.93	45.17	9.91	76.01
1800mm	nr	1.50	23.25	65.72	13.35	102.32
2000mm	nr	1.60	24.80	87.38	16.83	129.01
700mm high, length						
500mm	nr	1.20	18.60	21.48	6.01	46.09
1000mm	nr	1.35	20.93	37.54	8.77	67.23
1400mm	nr	1.45	22.48	69.04	13.73	105.24
1800mm	nr	1.60	24.80	89.59	17.16	131.55
2000mm	nr	1.70	26.35	99.32	18.85	144.52

	Unit	Labour hours	Hours cost £	Materials £	O & P £	Total £
Double convector radiator fixed to concealed brackets plugged and screwed to brickwork						
300mm high, length						
400mm	nr	0.90	13.95	26.85	6.12	46.92
800mm	nr	1.05	16.28	47.18	9.52	72.97
1200mm	nr	1.15	17.83	67.53	12.80	98.16
1600mm	nr	1.30	20.15	91.11	16.69	127.95
2000mm	nr	1.40	21.70	112.54	20.14	154.38
450mm high, length						
500mm	nr	1.00	15.50	25.42	6.14	47.06
1000mm	nr	1.15	17.83	46.86	9.70	74.39
1400mm	nr	1.25	19.38	72.28	13.75	105.40
1800mm	nr	1.40	21.70	116.67	20.76	159.13
2000mm	nr	1.50	23.25	128.63	22.78	174.66
600mm high, length						
500mm	nr	1.10	17.05	30.48	7.13	54.66
1000mm	nr	1.25	19.38	57.71	11.56	88.65
1400mm	nr	1.35	20.93	89.70	16.59	127.22
1800mm	nr	1.50	23.25	143.95	25.08	192.28
2000mm	nr	1.60	24.80	160.53	27.80	213.13
700mm high, length						
500mm	nr	1.20	18.60	34.58	7.98	61.16
1000mm	nr	1.35	20.93	65.75	13.00	99.68
1400mm	nr	1.45	22.48	122.21	21.70	166.39
1800mm	nr	1.60	24.80	163.37	28.23	216.40
2000mm	nr	1.70	26.35	179.61	30.89	236.85

	Unit	Labour hours	Hours cost £	Materials £	O & P £	Total £
Double panel radiator fixed to concealed brackets plugged and screwed to brickwork						
300mm high, length						
400mm	nr	0.90	13.95	21.42	5.31	40.68
800mm	nr	1.05	16.28	37.42	8.05	61.75
1200mm	nr	1.15	17.83	54.51	10.85	83.19
1600mm	nr	1.30	20.15	71.58	13.76	105.49
2000mm	nr	1.40	21.70	87.59	16.39	125.68
450mm high, length						
500mm	nr	1.00	15.50	21.08	5.49	42.07
1000mm	nr	1.15	17.83	37.09	8.24	63.15
1400mm	nr	1.25	19.38	57.09	11.47	87.93
1800mm	nr	1.40	21.70	91.71	17.01	130.42
2000mm	nr	1.50	23.25	101.51	18.71	143.47
600mm high, length						
500mm	nr	1.10	17.05	25.06	6.32	48.43
1000mm	nr	1.25	19.38	45.77	9.77	74.92
1400mm	nr	1.35	20.93	71.26	13.83	106.01
1800mm	nr	1.50	23.25	114.22	20.62	158.09
2000mm	nr	1.60	24.80	126.90	22.76	174.46
700mm high, length						
500mm	nr	1.20	18.60	32.33	7.64	58.57
1000mm	nr	1.35	20.93	62.49	12.51	95.93
1400mm	nr	1.45	22.48	97.25	17.96	137.68
1800mm	nr	1.60	24.80	129.74	23.18	177.72
2000mm	nr	1.70	26.35	143.80	25.52	195.67

	Unit	Labour hours	Hours cost £	Materials £	O & P £	Total £
ALTERATIONS AND REPAIRS						
Rainwater goods						
Remove existing length of gutter and fittings and prepare to receive new						
cast iron						
up to 100mm	nr	0.50	7.75	0.00	1.16	8.91
over 100mm	nr	0.60	9.30	0.00	1.40	10.70
PVC-U						
up to 100mm	nr	0.30	4.65	0.00	0.70	5.35
over 100mm	nr	0.35	5.43	0.00	0.81	6.24
aluminium						
up to 100mm	nr	0.50	7.75	0.00	1.16	8.91
over 100mm	nr	0.60	9.30	0.00	1.40	10.70
Remove single length of gutter and fittings and fix new to existing brackets						
cast iron						
100mm	nr	1.30	20.15	23.65	6.57	50.37
115mm	nr	1.35	20.93	24.89	6.87	52.69
125mm	nr	1.40	21.70	27.61	7.40	56.71
150-mm	nr	1.45	22.48	47.68	10.52	80.68
PVC-U						
75mm	nr	0.90	13.95	13.69	4.15	31.79
110mm	nr	0.95	14.73	11.02	3.86	29.61
160mm	nr	1.00	15.50	14.61	4.52	34.63
aluminium						
100mm	nr	0.90	13.95	27.81	6.26	48.02
125mm	nr	0.95	14.73	32.15	7.03	53.91

	Unit	Labour hours	Hours cost £	Materials £	O & P £	Total £
Remove gutter fittings, prepare ends, and fix new fitting						
cast iron, 100mm						
angle	nr	0.50	7.75	8.77	2.48	19.00
stop end outlet	nr	0.40	6.20	9.04	2.29	17.53
cast iron, 115mm						
angle	nr	0.50	7.75	9.14	2.53	19.42
stop end outlet	nr	0.40	6.20	9.98	2.43	18.61
cast iron, 125mm						
angle	nr	0.50	7.75	9.53	2.59	19.87
stop end outlet	nr	0.40	6.20	11.57	2.67	20.44
cast iron, 150mm						
angle	nr	0.50	7.75	17.32	3.76	28.83
stop end outlet	nr	0.40	6.20	18.77	3.75	28.72
PVC-U, 76mm						
angle	nr	0.50	7.75	6.54	2.14	16.43
stop end outlet	nr	0.40	6.20	6.03	1.83	14.06
PVC-U, 112mm						
angle	nr	0.50	7.75	6.88	2.19	16.82
stop end outlet	nr	0.40	6.20	6.28	1.87	14.35
PVC-U, 150mm						
angle	nr	0.50	7.75	9.98	2.66	20.39
stop end outlet	nr	0.40	6.20	10.22	2.46	18.88
aluminium, 100mm						
angle	nr	0.50	7.75	8.06	2.37	18.18
stop end outlet	nr	0.40	6.20	7.66	2.08	15.94
aluminium, 125mm						
angle	nr	0.50	7.75	8.98	2.51	19.24
stop end outlet	nr	0.40	6.20	8.36	2.18	16.74

	Unit	Labour hours	Hours cost £	Materials £	O & P £	Total £
Remove existing gutter brackets and replace with galvanised steel repair brackets and 1m maximum centres						
cast iron						
100mm	nr	0.25	3.88	2.69	0.98	7.55
115mm	nr	0.27	4.19	2.69	1.03	7.91
125mm	nr	0.30	4.65	2.69	1.10	8.44
PVC-U						
76mm	nr	0.25	3.88	2.69	0.98	7.55
112mm	nr	0.27	4.19	2.69	1.03	7.91
160mm	nr	0.30	4.65	2.69	1.10	8.44
aluminium						
100mm	nr	0.25	3.88	2.69	0.98	7.55
125mm	nr	0.30	4.65	2.69	1.10	8.44
Remove pipes and fittings and prepare to receive new pipework						
cast iron						
up to 100mm	m	0.40	6.20	0.00	0.93	7.13
over 100mm	m	0.50	7.75	0.00	1.16	8.91
PVC-U						
up to 100mm	m	0.30	4.65	0.00	0.70	5.35
over 100mm	m	0.40	6.20	0.00	0.93	7.13
aluminium						
up to 100mm	m	0.30	4.65	0.00	0.70	5.35
over 100mm	m	0.40	6.20	0.00	0.93	7.13

	Unit	Labour hours	Hours cost £	Materials £	O & P £	Total £
Remove length of pipe and replace with new						
cast iron						
65mm diameter	nr	0.70	10.85	0.00	1.63	12.48
75mm diameter	nr	0.70	10.85	0.00	1.63	12.48
100mm diameter	nr	0.80	12.40	0.00	1.86	14.26
PVC-U						
68mm diameter	nr	0.50	7.75	0.00	1.16	8.91
68mm square	nr	0.60	9.30	0.00	1.40	10.70
aluminium						
63mm diameter	nr	0.50	7.75	0.00	1.16	8.91
76mm diameter	nr	0.50	7.75	0.00	1.16	8.91
102mm diameter	nr	0.60	9.30	0.00	1.40	10.70
Remove existing pipe fittings and replace with new						
cast iron offset						
65mm diameter	nr	0.28	4.34	16.67	3.15	24.16
75mm diameter	nr	0.30	4.65	16.67	3.20	24.52
100mm diameter	nr	0.32	4.96	33.25	5.73	43.94
cast iron shoe						
65mm diameter	nr	0.28	4.34	16.35	3.10	23.79
75mm diameter	nr	0.30	4.65	16.35	3.15	24.15
100mm diameter	nr	0.32	4.96	24.67	4.44	34.07
cast iron shoe, eared						
65mm diameter	nr	0.30	4.65	18.67	3.50	26.82
75mm diameter	nr	0.32	4.96	18.67	3.54	27.17
100mm diameter	nr	0.34	5.27	25.71	4.65	35.63

	Unit	Labour hours	Hours cost £	Materials £	O & P £	Total £
cast iron bend						
65mm diameter	nr	0.24	3.72	11.48	2.28	17.48
75mm diameter	nr	0.32	4.96	11.48	2.47	18.91
100mm diameter	nr	0.24	3.72	19.64	3.50	26.86
PVC-U bend						
68mm diameter	nr	0.20	3.10	6.01	1.37	10.48
68mm square	nr	0.20	3.10	6.98	1.51	11.59
PVC-U offset						
68mm diameter	nr	0.20	3.10	4.06	1.07	8.23
68mm square	nr	0.20	3.10	6.25	1.40	10.75
PVC-U shoe						
68mm diameter	nr	0.20	3.10	5.87	1.35	10.32
68mm square	nr	0.20	3.10	6.34	1.42	10.86
aluminium offset						
63mm diameter	nr	0.22	3.41	34.64	5.71	43.76
76mm diameter	nr	0.24	3.72	46.36	7.51	57.59
102mm diameter	nr	0.26	4.03	52.22	8.44	64.69
aluminium bend						
63mm diameter	nr	0.22	3.41	27.55	4.64	35.60
76mm diameter	nr	0.24	3.72	32.67	5.46	41.85
102mm diameter	nr	0.26	4.03	43.08	7.07	54.18
Rake out existing cement mortar joint of soil pipe connection and re-point						
65mm diameter	nr	0.30	4.65	0.60	0.79	6.04
75mm diameter	nr	0.30	4.65	0.65	0.80	6.10
100mm diameter	nr	0.30	4.65	0.70	0.80	6.15

	Unit	Labour hours	Hours cost £	Materials £	O & P £	Total £
Hot and cold water supply						
Cut out 500mm length of copper pipe,fix new length with compression fittings each end						
15mm	nr	0.30	4.65	4.28	1.34	10.27
18mm	nr	0.30	4.65	5.88	1.58	12.11
22mm	nr	0.30	4.65	6.27	1.64	12.56
28mm	nr	0.30	4.65	8.99	2.05	15.69
35mm	nr	0.30	4.65	17.05	3.26	24.96
42mm	nr	0.30	4.65	19.99	3.70	28.34
54mm	nr	0.30	4.65	24.66	4.40	33.71
Take off existing radiator valve, replace with new, drain down system and bleed after installation						
single valve	nr	1.00	15.50	18.97	5.17	39.64
complete system, 9 valves	nr	6.00	93.00	195.56	43.28	331.84
Take out existing steel water storage tank, fix new plastic tank complete with valve, lid and insulation						
18 litres tank	nr	4.00	62.00	26.35	13.25	101.60
68 litres tank	nr	4.50	69.75	33.51	15.49	118.75
154 litres tank	nr	5.00	77.50	54.66	19.82	151.98
Sanitary fittings						
Take out high-level WC, instal new low-level WC suite complete	nr	5.00	77.50	305.66	57.47	440.63

	Unit	Labour hours	Hours cost £	Materials £	O & P £	Total £
Take out cast iron bath instal new acrylic bath complete with panels, taps, shower and handset	nr	4.50	69.75	397.54	70.09	537.38
Take out wash basin, instal new basin complete	nr	4.00	62.00	143.62	30.84	236.46

Part Two

PROJECT COSTS

Rainwater goods

Bathrooms

External waste systems

Central heating systems

Hot and cold water

	Qty	Unit	Labour hours	Hours cost £	Materials £	O & P £	Total £
PVC-U RAINWATER GOODS							
One-storey terraced house size 7 x 6m with gable ends							
Take down existing rainwater goods and make good		Item	4.00	62.00	0.00	9.30	71.30
68mm diameter PVC-U rainwater pipe with pipe clips plugged to brickwork at 2m maximum centres	6	m	1.50	23.25	36.48	8.96	68.69
extra over for							
branch	2	nr	0.50	7.75	11.68	2.91	22.34
shoe	2	nr	0.50	7.75	10.08	2.67	20.50
110mm half round PVC-U gutter to softwood fascia	14	m	2.80	43.40	62.72	15.92	122.04
extra over for							
running outlet	2	nr	0.52	8.06	8.26	2.45	18.77
Total			9.82	152.21	129.22	42.21	323.64
One-storey semi-detached house size 8 x 7m with gable ends							
Take down existing rainwater goods and make good		Item	6.00	93.00	0.00	13.95	106.95
68mm diameter PVC-U rainwater pipe with pipe clips plugged to brickwork at 2m maximum centres	6	m	1.50	23.25	36.48	8.96	68.69
Carried forward			7.50	116.25	36.48	22.91	175.64

	Qty	Unit	Labour hours	Hours cost £	Materials £	O & P £	Total £
Brought forward			7.50	116.25	36.48	22.91	175.64
extra over for							
offset	2	nr	0.30	4.65	11.68	2.45	18.78
shoe	2	nr	0.60	9.30	10.08	2.91	22.29
110mm half round PVC-U gutter to softwood fascia	16	m	3.20	49.60	71.68	18.19	139.47
extra over for							
stop end outlet	2	nr	0.26	4.03	8.26	1.84	14.13
Total			11.86	183.83	138.18	48.30	370.31

	Qty	Unit	Labour hours	Hours cost £	Materials £	O & P £	Total £
One-storey detached house size 9 x 8m with gable ends							
Take down existing rainwater goods and make good		Item	7.00	108.50	0.00	16.28	124.78
68mm diameter PVC-U rainwater pipe with pipe clips plugged to brickwork at 2m maximum centres	6	m	1.50	23.25	36.48	8.96	68.69
extra over for							
offset	2	nr	0.30	4.65	11.68	2.45	18.78
shoe	2	nr	0.60	9.30	10.08	2.91	22.29
110mm half round PVC-U gutter to softwood fascia	18	m	3.60	55.80	80.64	20.47	156.91
extra over for							
running outlet	2	nr	0.26	4.03	8.26	1.84	14.13
Total			13.26	205.53	147.14	52.90	405.57

	Qty	Unit	Labour hours	Hours cost £	Materials £	O & P £	Total £
Two-storey terraced house size 7 x 7m with gable ends							
Take down existing rainwater goods and make good		Item	5.00	77.50	0.00	11.63	89.13
68mm diameter PVC-U rainwater pipe with pipe clips plugged to brickwork at 2m maximum centres	12	m	3.00	46.50	72.96	17.92	137.38
extra over for							
offset	2	nr	0.50	7.75	11.68	2.91	22.34
shoe	2	nr	0.50	7.75	10.08	2.67	20.50
110mm half round PVC-U gutter to softwood fascia	14	m	2.80	43.40	62.72	15.92	122.04
extra over for							
running outlet	2	nr	0.52	8.06	8.26	2.45	18.77
Total			12.32	190.96	165.70	53.50	410.16

	Qty	Unit	Labour hours	Hours cost £	Materials £	O & P £	Total £
Two-storey semi-detached house size 8 x 7m with gable ends							
Take down existing rainwater goods and make good		Item	7.00	108.50	0.00	16.28	124.78
68mm diameter PVC-U rainwater pipe with pipe clips plugged to brickwork at 2m maximum centres	12	m	3.00	46.50	72.96	17.92	137.38
Carried forward			10.00	155.00	72.96	34.19	262.15

	Qty	Unit	Labour hours	Hours cost £	Materials £	O & P £	Total £
Brought forward			10.00	155.00	72.96	34.19	262.15
extra over for							
offset	2	nr	0.50	7.75	11.68	2.91	22.34
shoe	2	nr	0.50	7.75	10.08	2.67	20.50
110mm half round PVC-U gutter to softwood fascia	16	m	3.60	55.80	71.68	19.12	146.60
extra over for							
stop end outlet	2	nr	0.26	4.03	8.26	1.84	14.13
Total			14.86	230.33	174.66	60.74	465.73

	Qty	Unit	Labour hours	Hours cost £	Materials £	O & P £	Total £
Two-storey detached house size 9 x 8m with gable ends							
Take down existing rainwater goods and make good		Item	8.00	116.00	0.00	17.40	133.40
68mm diameter PVC-U rainwater pipe with pipe clips plugged to brickwork at 2m maximum centres	12	m	3.00	46.50	72.96	17.92	137.38
extra over for							
offset	2	nr	0.50	7.75	11.68	2.91	22.34
shoe	2	nr	0.50	7.75	10.08	2.67	20.50
110mm half round PVC-U gutter to softwood fascia	18	m	3.60	55.80	80.64	20.47	156.91
extra over for							
stop end outlet	2	nr	0.26	4.03	8.26	1.84	14.13
Total			15.86	237.83	183.62	63.22	484.67

	Qty	Unit	Labour hours	Hours cost £	Materials £	O & P £	Total £
Three-storey terraced house size 8 x 8m with gable ends							
Take down existing rainwater goods and make good		Item	6.00	93.00	0.00	13.95	106.95
68mm diameter PVC-U rainwater pipe with pipe clips plugged to brickwork at 2m maximum centres	18	m	4.50	69.75	109.44	26.88	206.07
extra over for							
offset	2	nr	0.50	7.75	11.68	2.91	22.34
shoe	2	nr	0.50	7.75	10.08	2.67	20.50
110mm half round PVC-U gutter to softwood fascia	16	m	3.20	49.60	71.68	18.19	139.47
extra over for							
running outlet	2	nr	0.52	8.06	8.26	2.45	18.77
Total			15.22	235.91	211.14	67.06	514.11

	Qty	Unit	Labour hours	Hours cost £	Materials £	O & P £	Total £
Three-storey semi-detached house size 9 x 8m with gable ends							
Take down existing rainwater goods and make good		Item	7.00	108.50	0.00	16.28	124.78
68mm diameter PVC-U rainwater pipe with pipe clips plugged to brickwork at 2m maximum centres	18	m	4.50	69.75	109.44	26.88	206.07
Carried forward			11.50	178.25	109.44	43.15	330.84

	Qty	Unit	Labour hours	Hours cost £	Materials £	O & P £	Total £
Brought forward			11.50	178.25	109.44	43.15	330.84
extra over for							
offset	2	nr	0.50	7.75	11.68	2.91	22.34
shoe	2	nr	0.50	7.75	10.08	2.67	20.50
110mm half round PVC-U gutter to softwood fascia	18	m	3.60	55.80	80.64	20.47	156.91
extra over for							
stop end outlet	2	nr	0.26	4.03	8.26	1.84	14.13
Total			16.36	253.58	220.10	71.05	544.73

	Qty	Unit	Labour hours	Hours cost £	Materials £	O & P £	Total £
Three-storey detached house size 10 x 9m with gable ends							
Take down existing rainwater goods and make good		Item	9.00	139.50	0.00	20.93	160.43
68mm diameter PVC-U rainwater pipe with pipe clips plugged to brickwork at 2m maximum centres	18	m	4.50	69.75	109.44	26.88	206.07
extra over for							
offset	2	nr	0.50	7.75	11.68	2.91	22.34
shoe	2	nr	0.50	7.75	10.08	2.67	20.50
110mm half round PVC-U gutter to softwood fascia	20	m	4.00	58.00	89.60	22.14	169.74
extra over for							
stop end outlet	2	nr	0.26	4.03	8.26	1.84	14.13
Total			18.76	286.78	229.06	77.38	593.22

	Qty	Unit	Labour hours	Hours cost £	Materials £	O & P £	Total £
One-storey terraced house size 7 x 6m with hipped ends							
Take down existing rainwater goods and make good		Item	4.00	62.00	0.00	9.30	71.30
68mm diameter PVC-U rainwater pipe with pipe clips plugged to brickwork at 2m maximum centres	6	m	1.50	23.25	36.48	8.96	68.69
extra over for							
offset	2	nr	0.50	7.75	11.68	2.91	22.34
shoe	2	nr	0.50	7.75	10.08	2.67	20.50
110mm half round PVC-U gutter to softwood fascia	14	m	2.80	43.40	62.72	15.92	122.04
extra over for							
running outlet	2	nr	0.52	8.06	8.26	2.45	18.77
Total			9.82	152.21	129.22	42.21	323.64
One-storey semi-detached house size 8 x 7m with hipped ends							
Take down existing rainwater goods and make good		Item	6.00	93.00	0.00	13.95	106.95
68mm diameter PVC-U rainwater pipe with pipe clips plugged to brickwork at 2m maximum centres	6	m	1.50	23.25	36.48	8.96	68.69
extra over for							
offset	2	nr	0.50	7.75	11.68	2.91	22.34
shoe	2	nr	0.50	7.75	10.08	2.67	20.50
Carried forward			8.50	131.75	58.24	28.50	218.49

	Qty	Unit	Labour hours	Hours cost £	Materials £	O & P £	Total £
Brought forward			8.50	131.75	58.24	28.50	218.49
110mm half round PVC-U gutter to softwood fascia	23	m	4.60	71.30	103.04	26.15	200.49
extra over for							
angle	2	nr	0.52	8.06	9.64	2.66	20.36
running outlet	2	nr	0.52	8.06	8.26	2.45	18.77
Total			14.14	219.17	179.18	59.75	458.10

	Qty	Unit	Labour hours	Hours cost £	Materials £	O & P £	Total £
One-storey detached house size 9 x 7m with hipped ends							
Take down existing rainwater goods and make good		Item	8.00	116.00	0.00	17.40	133.40
68mm diameter PVC-U rainwater pipe with pipe clips plugged to brickwork at 2m maximum centres	6	m	1.50	23.25	36.48	8.96	68.69
extra over for							
offset	2	nr	0.50	7.75	11.68	2.91	22.34
shoe	2	nr	0.50	7.75	10.08	2.67	20.50
110mm half round PVC-U gutter to softwood fascia	32	m	6.40	99.20	143.46	36.40	279.06
extra over for							
angle	4	nr	0.52	8.06	9.64	2.66	20.36
running outlet	2	nr	1.04	16.12	8.26	3.66	28.04
Total			18.46	278.13	219.60	74.66	572.39

	Qty	Unit	Labour hours	Hours cost £	Materials £	O & P £	Total £
Two-storey terraced house size 7 x 7m with hipped ends							
Take down existing rainwater goods and make good		Item	6.00	93.00	0.00	13.95	106.95
68mm diameter PVC-U rainwater pipe with pipe clips plugged to brickwork at 2m maximum centres	12	m	3.00	46.50	72.96	17.92	137.38
extra over for							
offset	2	nr	0.50	7.75	11.68	2.91	22.34
shoe	2	nr	0.50	7.75	10.08	2.67	20.50
110mm half round PVC-U gutter to softwood fascia	14	m	2.80	43.40	62.72	15.92	122.04
extra over for							
running outlet	2	nr	0.52	8.06	8.26	2.45	18.77
Total			13.32	206.46	165.70	55.82	427.98

	Qty	Unit	Labour hours	Hours cost £	Materials £	O & P £	Total £
Two-storey semi-detached house size 8 x 7m with hipped ends							
Take down existing rainwater goods and make good		Item	8.00	124.00	0.00	18.60	142.60
68mm diameter PVC-U rainwater pipe with pipe clips plugged to brickwork at 2m maximum centres	12	m	3.00	46.50	72.96	17.92	137.38
Carried forward			11.00	170.50	72.96	36.52	279.98

	Qty	Unit	Labour hours	Hours cost £	Materials £	O & P £	Total £
Brought forward			11.00	170.50	72.96	36.52	279.98
extra over for							
offset	2	nr	0.50	7.75	11.68	2.91	22.34
shoe	2	nr	0.50	7.75	10.08	2.67	20.50
110mm half round PVC-U gutter to softwood fascia	24	m	4.60	71.30	103.04	26.15	200.49
extra over for							
angle	2	nr	0.52	8.06	9.64	2.66	20.36
running outlet	2	nr	0.52	8.06	8.26	2.45	18.77
Total			17.64	273.42	215.66	73.36	562.44

	Qty	Unit	Labour hours	Hours cost £	Materials £	O & P £	Total £
Two-storey detached house size 9 x 8m with hipped ends							
Take down existing rainwater goods and make good		Item	9.00	130.50	0.00	19.58	150.08
68mm diameter PVC-U rainwater pipe as before	12	m	3.00	46.50	72.26	17.81	136.57
extra over for							
offset	2	nr	0.50	7.75	11.68	2.91	22.34
shoe	2	nr	0.50	7.75	10.08	2.67	20.50
110mm half round PVC-U gutter to softwood fascia	34	m	6.80	105.40	143.36	37.31	286.07
extra over for							
angle	4	nr	0.52	8.06	9.64	2.66	20.36
running outlet	2	nr	1.04	16.12	19.28	5.31	40.71
Total			21.36	322.08	266.30	88.26	676.64

	Qty	Unit	Labour hours	Hours cost £	Materials £	O & P £	Total £
Three-storey terraced house size 8 x 8m with hipped ends							
Take down existing rainwater goods and make good		Item	7.00	108.50	0.00	16.28	124.78
68mm diameter PVC-U rainwater pipe with pipe clips plugged to brickwork at 2m maximum centres	18	m	4.50	69.75	109.44	26.88	206.07
extra over for							
offset	2	nr	0.50	7.75	11.68	2.91	22.34
shoe	2	nr	0.50	7.75	10.08	2.67	20.50
110mm half round PVC-U gutter to softwood fascia	16	m	3.20	49.60	71.68	18.19	139.47
extra over for							
running outlet	2	nr	0.52	8.06	8.26	2.45	18.77
Total			16.22	251.41	211.14	69.38	531.93

	Qty	Unit	Labour hours	Hours cost £	Materials £	O & P £	Total £
Three-storey semi-detached house size 9 x 8m with hipped ends							
Take down existing rainwater goods and make good		Item	9.00	139.50	0.00	20.93	160.43
68mm diameter PVC-U rainwater pipe with pipe clips plugged to brickwork at 2m maximum centres	18	m	4.50	69.75	109.44	26.88	206.07
Carried forward			13.50	209.25	109.44	47.80	366.49

	Qty	Unit	Labour hours	Hours cost £	Materials £	O & P £	Total £
Brought forward			13.50	209.25	109.44	47.80	366.49
extra over for							
offset	2	nr	0.50	7.75	11.68	2.91	22.34
shoe	2	nr	0.50	7.75	10.08	2.67	20.50
110mm half round PVC-U gutter to softwood fascia	26	m	5.20	80.60	116.48	29.56	226.64
extra over for							
angle	2	nr	0.52	8.06	9.64	2.66	20.36
running outlet	2	nr	0.52	8.06	8.26	2.45	18.77
Total			20.74	321.47	265.58	88.06	675.11

	Qty	Unit	Labour hours	Hours cost £	Materials £	O & P £	Total £
Three-storey detached house size 10 x 9m with hipped ends							
Take down existing rainwater goods and make good		Item	10.00	155.00	0.00	23.25	178.25
68mm diameter PVC-U rainwater pipe as before	18	m	4.50	69.75	109.44	26.88	206.07
extra over for							
offset	2	nr	0.50	7.75	11.68	2.91	22.34
shoe	2	nr	0.50	7.75	10.08	2.67	20.50
110mm half round PVC-U gutter to softwood fascia	38	m	7.60	117.80	170.24	43.21	331.25
extra over for							
angle	4	nr	0.52	8.06	8.26	2.45	18.77
running outlet	2	nr	0.52	8.06	9.64	2.66	20.36
Total			24.14	374.17	319.34	104.03	797.54

	Qty	Unit	Labour hours	Hours cost £	Materials £	O & P £	Total £
CAST IRON RAINWATER GOODS							
One-storey terraced house size 7 x 6m with gable ends							
Take down existing rainwater goods and make good		Item	4.00	62.00	0.00	9.30	71.30
75mm diameter cast iron rainwater pipe with pipe clips plugged to brickwork at 2m maximum centres	6	m	1.80	27.90	133.50	24.21	185.61
extra over for							
offset	2	nr	0.60	9.30	33.00	6.35	48.65
shoe	2	nr	0.60	9.30	30.68	6.00	45.98
100mm half round cast iron gutter to softwood fascia	14	m	4.20	65.10	196.70	39.27	301.07
extra over for							
running outlet	2	nr	0.60	9.30	15.98	3.79	29.07
Total			11.80	182.90	409.86	88.91	681.67
One-storey semi-detached house size 8 x 7m with gable ends							
Take down existing rainwater goods and make good		Item	6.00	93.00	0.00	13.95	106.95
75mm diameter cast iron rainwater pipe with pipe clips plugged to brickwork at 2m maximum centres	6	m	1.80	27.90	133.50	24.21	185.61
Carried forward			7.80	120.90	133.50	38.16	292.56

	Qty	Unit	Labour hours	Hours cost £	Materials £	O & P £	Total £
Brought forward			7.80	120.90	133.50	38.16	292.56
extra over for							
offset	2	nr	0.60	9.30	33.00	6.35	48.65
shoe	2	nr	0.60	9.30	30.68	6.00	45.98
100mm half round cast iron gutter to softwood fascia	16	m	4.80	74.40	224.80	44.88	344.08
extra over for							
stop end outlet	2	nr	0.30	4.65	15.98	3.09	23.72
Total			14.10	218.55	437.96	98.48	754.99

	Qty	Unit	Labour hours	Hours cost £	Materials £	O & P £	Total £
One-storey detached house size 9 x 8m with gable ends							
Take down existing rainwater goods and make good		Item	7.00	108.50	0.00	16.28	124.78
75mm diameter cast iron rainwater pipe with pipe clips plugged to brickwork at 2m maximum centres	6	m	1.80	27.90	133.50	24.21	185.61
extra over for							
offset	2	nr	0.60	9.30	33.00	6.35	48.65
shoe	2	nr	0.60	9.30	30.68	6.00	45.98
100mm half round cast iron gutter to softwood fascia	18	m	5.40	83.70	252.90	50.49	387.09
extra over for							
running outlet	2	nr	0.30	4.65	19.88	3.68	28.21
Total			15.70	243.35	469.96	107.00	820.31

	Qty	Unit	Labour hours	Hours cost £	Materials £	O & P £	Total £
Two-storey terraced house size 7 x 7m with gable ends							
Take down existing rainwater goods and make good		Item	5.00	77.50	0.00	11.63	89.13
75mm diameter cast iron rainwater pipe with pipe clips plugged to brickwork at 2m maximum centres	12	m	3.60	55.80	267.00	48.42	371.22
extra over for							
offset	2	nr	0.60	9.30	33.00	6.35	48.65
shoe	2	nr	0.60	9.30	30.68	6.00	45.98
100mm half round cast iron gutter to softwood fascia	14	m	4.20	65.10	196.70	39.27	301.07
extra over for							
running outlet	2	nr	0.60	9.30	15.98	3.79	29.07
Total			14.60	226.30	543.36	115.45	885.11

	Qty	Unit	Labour hours	Hours cost £	Materials £	O & P £	Total £
Two-storey semi-detached house size 8 x 7m with gable ends							
Take down existing rainwater goods and make good		Item	7.00	108.50	0.00	16.28	124.78
75mm diameter cast iron rainwater pipe with pipe clips plugged to brickwork at 2m maximum centres	12	m	3.60	55.80	267.00	48.42	371.22
Carried forward			10.60	164.30	267.00	64.70	496.00

	Qty	Unit	Labour hours	Hours cost £	Materials £	O & P £	Total £
Brought forward			10.60	164.30	267.00	64.70	496.00
extra over for							
offset	2	nr	0.60	9.30	33.00	6.35	48.65
shoe	2	nr	0.60	9.30	30.68	6.00	45.98
100mm half round cast iron gutter to softwood fascia	16	m	4.80	74.40	224.80	44.88	344.08
extra over for							
stop end outlet	2	nr	0.30	4.65	19.88	3.68	28.21
Total			16.90	261.95	575.36	125.60	962.91

	Qty	Unit	Labour hours	Hours cost £	Materials £	O & P £	Total £
Two-storey detached house size 9 x 8m with gable ends							
Take down existing rainwater goods and make good		Item	8.00	124.00	0.00	18.60	142.60
75mm diameter cast iron rainwater pipe with pipe clips plugged to brickwork at 2m maximum centres	12	m	3.60	55.80	267.00	48.42	371.22
extra over for							
offset	2	nr	0.60	9.30	33.00	6.35	48.65
shoe	2	nr	0.60	9.30	30.68	6.00	45.98
100mm half round cast iron gutter to softwood fascia	18	m	5.40	83.70	252.90	50.49	387.09
extra over for							
stop end outlet	2	nr	0.30	4.65	19.88	3.68	28.21
Total			18.50	286.75	603.46	133.53	1,023.74

	Qty	Unit	Labour hours	Hours cost £	Materials £	O & P £	Total £
Three-storey terraced house size 8 x 8m with gable ends							
Take down existing rainwater goods and make good		Item	6.00	93.00	0.00	13.95	106.95
75mm diameter cast iron rainwater pipe with pipe clips plugged to brickwork at 2m maximum centres	18	m	5.40	83.70	400.50	72.63	556.83
extra over for							
offset	2	nr	0.60	9.30	33.00	6.35	48.65
shoe	2	nr	0.60	9.30	30.68	6.00	45.98
100mm half round cast iron gutter to softwood fascia	16	m	1.80	27.90	196.70	33.69	258.29
extra over for							
running outlet	2	nr	0.60	9.30	25.98	5.29	40.57
Total			15.00	232.50	686.86	137.90	1,057.26

	Qty	Unit	Labour hours	Hours cost £	Materials £	O & P £	Total £
Three-storey semi-detached house size 9 x 8m with gable ends							
Take down existing rainwater goods and make good		Item	7.00	108.50	0.00	16.28	124.78
75mm diameter cast iron rainwater pipe with pipe clips plugged to brickwork at 2m maximum centres	18	m	5.40	83.70	400.50	72.63	556.83
Carried forward			12.40	192.20	400.50	88.91	681.61

	Qty	Unit	Labour hours	Hours cost £	Materials £	O & P £	Total £
Brought forward			12.40	192.20	400.50	88.91	681.61
extra over for							
offset	2	nr	0.60	9.30	33.00	6.35	48.65
shoe	2	nr	0.60	9.30	30.68	6.00	45.98
100mm half round cast iron gutter to softwood fascia	18	m	5.40	83.70	224.80	46.28	354.78
extra over for							
stop end outlet	2	nr	0.30	4.65	19.88	3.68	28.21
Total			19.30	299.15	708.86	151.20	1,159.21

	Qty	Unit	Labour hours	Hours cost £	Materials £	O & P £	Total £
Three-storey detached house size 10 x 9m with gable ends							
Take down existing rainwater goods and make good		Item	9.00	139.50	0.00	20.93	160.43
75mm diameter cast iron rainwater pipe with pipe clips plugged to brickwork at 2m maximum centres	18	m	5.40	83.70	400.50	72.63	556.83
extra over for							
offset	2	nr	0.60	9.30	33.00	6.35	48.65
shoe	2	nr	0.60	9.30	30.68	6.00	45.98
100mm half round cast iron gutter to softwood fascia	20	m	6.00	93.00	252.90	51.89	397.79
extra over for							
stop end outlet	2	nr	0.30	4.65	19.88	3.68	28.21
Total			21.90	339.45	736.96	161.46	1,237.87

	Qty	Unit	Labour hours	Hours cost £	Materials £	O & P £	Total £
One-storey terraced house size 7 x 6m with hipped ends							
Take down existing rainwater goods and make good		Item	4.00	62.00	0.00	9.30	71.30
75mm diameter cast iron rainwater pipe with pipe clips plugged to brickwork at 2m maximum centres	6	m	1.80	27.90	133.50	24.21	185.61
extra over for							
offset	2	nr	0.60	9.30	33.04	6.35	48.69
shoe	2	nr	0.60	9.30	30.68	6.00	45.98
100mm half round cast iron gutter to softwood fascia	14	m	4.20	65.10	196.70	39.27	301.07
extra over for							
running outlet	2	nr	0.60	9.30	15.98	3.79	29.07
Total			11.80	182.90	409.90	88.92	681.72
One-storey semi-detached house size 8 x 7m with hipped ends							
Take down existing rainwater goods and make good		Item	6.00	93.00	0.00	13.95	106.95
75mm diameter cast iron rainwater pipe with pipe clips plugged to brickwork at 2m maximum centres	6	m	1.80	27.90	133.50	24.21	185.61
extra over for							
offset	2	nr	0.60	9.30	33.00	6.35	48.65
shoe	2	nr	0.60	9.30	30.68	6.00	45.98
Carried forward			9.00	139.50	197.18	50.50	387.18

	Qty	Unit	Labour hours	Hours cost £	Materials £	O & P £	Total £
Brought forward			9.00	139.50	197.18	50.50	387.18
100mm half round cast iron gutter to softwood fascia	23	m	6.90	106.95	226.80	50.06	383.81
extra over for							
angle	2	nr	0.60	9.30	15.98	3.79	29.07
running outlet	2	nr	0.60	9.30	19.88	4.38	33.56
Total			17.10	265.05	459.84	108.73	833.62

	Qty	Unit	Labour hours	Hours cost £	Materials £	O & P £	Total £
One-storey detached house size 9 x 7m with hipped ends							
Take down existing rainwater goods and make good		Item	8.00	124.00	0.00	18.60	142.60
75mm diameter cast iron rainwater pipe with pipe clips plugged to brickwork at 2m maximum centres	6	m	1.80	27.90	133.50	24.21	185.61
extra over for							
offset	2	nr	0.60	9.30	33.00	6.35	48.65
shoe	2	nr	0.60	9.30	30.68	6.00	45.98
100mm half round cast iron gutter to softwood fascia	32	m	9.60	148.80	449.60	89.76	688.16
extra over for							
angle	4	nr	0.60	9.30	39.68	7.35	56.33
running outlet	2	nr	1.20	18.60	15.98	5.19	39.77
Total			22.40	347.20	702.44	157.45	1,207.09

	Qty	Unit	Labour hours	Hours cost £	Materials £	O & P £	Total £
Two-storey terraced house size 7 x 7m with hipped ends							
Take down existing rainwater goods and make good		Item	6.00	93.00	0.00	13.95	106.95
75mm diameter cast iron rainwater pipe with pipe clips plugged to brickwork at 2m maximum centres	12	m	3.60	55.80	267.00	48.42	371.22
extra over for							
offset	2	nr	0.60	9.30	33.00	6.35	48.65
shoe	2	nr	0.60	9.30	30.68	6.00	45.98
100mm half round cast iron gutter to softwood fascia	14	m	4.20	65.10	196.70	39.27	301.07
extra over for							
running outlet	2	nr	0.60	9.30	15.98	3.79	29.07
Total			15.60	241.80	543.36	117.77	902.93

	Qty	Unit	Labour hours	Hours cost £	Materials £	O & P £	Total £
Two-storey semi-detached house size 8 x 7m with hipped ends							
Take down existing rainwater goods and make good		Item	8.00	124.00	0.00	18.60	142.60
75mm diameter cast iron rainwater pipe with pipe clips plugged to brickwork at 2m maximum centres	12	m	3.60	55.80	267.00	48.42	371.22
Carried forward			11.60	179.80	267.00	67.02	513.82

	Qty	Unit	Labour hours	Hours cost £	Materials £	O & P £	Total £
Brought forward			11.60	179.80	267.00	67.02	513.82
extra over for							
offset	2	nr	0.60	9.30	33.00	6.35	48.65
shoe	2	nr	0.60	9.30	30.68	6.00	45.98
100mm half round cast iron gutter to softwood fascia	23	m	6.90	106.95	323.15	64.52	494.62
extra over for							
angle	2	nr	0.60	9.30	19.84	4.37	33.51
running outlet	2	nr	0.60	9.30	15.98	3.79	29.07
Total			20.90	323.95	689.65	152.04	1,165.64

	Qty	Unit	Labour hours	Hours cost £	Materials £	O & P £	Total £
Two-storey detached house size 9 x 8m with hipped ends							
Take down existing rainwater goods and make good		Item	9.00	139.50	0.00	20.93	160.43
75mm diameter cast iron rainwater pipe as before	12	m	3.60	55.80	267.00	48.42	371.22
extra over for							
offset	2	nr	0.60	9.30	33.00	6.35	48.65
shoe	2	nr	0.60	9.30	30.68	6.00	45.98
100mm half round cast iron gutter to softwood fascia	34	m	10.20	158.10	477.70	95.37	731.17
extra over for							
angle	4	nr	0.60	9.30	39.68	7.35	56.33
running outlet	2	nr	1.20	18.60	15.98	5.19	39.77
Total			25.80	399.90	864.04	189.59	1,453.53

	Qty	Unit	Labour hours	Hours cost £	Materials £	O & P £	Total £
Three-storey terraced house size 8 x 8m with hipped ends							
Take down existing rainwater goods and make good		Item	7.00	108.50	0.00	16.28	124.78
75mm diameter cast iron rainwater pipe with pipe clips plugged to brickwork at 2m maximum centres	18	m	5.40	83.70	400.50	72.63	556.83
extra over for							
offset	2	nr	0.60	9.30	33.04	6.35	48.69
shoe	2	nr	0.60	9.30	30.68	6.00	45.98
100mm half round cast iron gutter to softwood fascia	16	m	4.80	74.40	224.80	44.88	344.08
extra over for							
running outlet	2	nr	0.60	9.30	15.98	3.79	29.07
Total			19.00	294.50	705.00	149.93	1,149.43

	Qty	Unit	Labour hours	Hours cost £	Materials £	O & P £	Total £
Three-storey semi-detached house size 9 x 8m with hipped ends							
Take down existing rainwater goods and make good		Item	9.00	139.50	0.00	20.93	160.43
75mm diameter cast iron rainwater pipe with pipe clips plugged to brickwork at 2m maximum centres	18	m	5.40	83.70	400.50	72.63	556.83
Carried forward			14.40	223.20	400.50	93.56	717.26

	Qty	Unit	Labour hours	Hours cost £	Materials £	O & P £	Total £
Brought forward			14.40	223.20	400.50	93.56	717.26
extra over for							
offset	2	nr	0.60	9.30	33.00	6.35	48.65
shoe	2	nr	0.60	9.30	30.68	6.00	45.98
100mm half round cast iron gutter to softwood fascia	26	m	7.80	120.90	365.30	72.93	559.13
extra over for							
angle	2	nr	0.60	9.30	15.98	3.79	29.07
running outlet	2	nr	0.60	9.30	19.84	4.37	33.51
Total			24.60	381.30	865.30	186.99	1,433.59

	Qty	Unit	Labour hours	Hours cost £	Materials £	O & P £	Total £
Three-storey detached house size 10 x 9m with hipped ends							
Take down existing rainwater goods and make good		Item	10.00	155.00	0.00	23.25	178.25
75mm diameter cast iron rainwater pipe as before	18	m	5.40	83.70	400.50	72.63	556.83
extra over for							
offset	2	nr	0.60	9.30	33.00	6.35	48.65
shoe	2	nr	0.60	9.30	30.68	6.00	45.98
100mm half round cast iron gutter to softwood fascia	38	m	11.40	176.70	533.90	106.59	817.19
extra over for							
angle	4	nr	0.60	9.30	15.98	3.79	29.07
running outlet	2	nr	1.20	18.60	39.68	8.74	67.02
Total			29.80	461.90	1053.74	227.35	1,742.99

	Qty	Unit	Labour hours	Hours cost £	Materials £	O & P £	Total £
ALUMINIUM RAINWATER GOODS							
One-storey terraced house size 7 x 6m with gable ends							
Take down existing rainwater goods and make good		Item	4.00	62.00	0.00	9.30	71.30
76mm diameter aluminium rainwater pipe with pipe clips plugged to brickwork at 2m maximum centres	6	m	1.80	27.90	122.94	22.63	173.47
extra over for							
offset	2	nr	0.60	9.30	57.58	10.03	76.91
shoe	2	nr	0.60	9.30	26.38	5.35	41.03
100mm half round aluminium gutter to softwood fascia	14	m	3.20	49.60	186.62	35.43	271.65
extra over for							
running outlet	2	nr	0.60	9.30	23.78	4.96	38.04
Total			10.80	167.40	417.30	87.71	672.41
One-storey semi-detached house size 8 x 7m with gable ends							
Take down existing rainwater goods and make good		Item	6.00	93.00	0.00	13.95	106.95
76mm diameter aluminium rainwater pipe with pipe clips plugged to brickwork at 2m maximum centres	6	m	1.80	27.90	122.94	22.63	173.47
Carried forward			7.80	120.90	122.94	36.58	280.42

	Qty	Unit	Labour hours	Hours cost £	Materials £	O & P £	Total £
Brought forward			7.80	120.90	122.94	36.58	280.42
extra over for							
offset	2	nr	0.60	9.30	57.58	10.03	76.91
shoe	2	nr	0.60	9.30	26.38	5.35	41.03
100mm half round aluminium gutter to softwood fascia	16	m	4.80	74.40	213.28	43.15	330.83
extra over for							
stop end outlet	2	nr	0.30	4.65	15.72	3.06	23.43
Total			14.10	218.55	435.90	98.17	752.62

	Qty	Unit	Labour hours	Hours cost £	Materials £	O & P £	Total £
One-storey detached house size 9 x 7m with gable ends							
Take down existing rainwater goods and make good		Item	7.00	108.50	0.00	16.28	124.78
76mm diameter aluminium rainwater pipe with pipe clips plugged to brickwork at 2m maximum centres	6	m	1.80	27.90	122.94	22.63	173.47
extra over for							
offset	2	nr	0.60	9.30	57.58	10.03	76.91
shoe	2	nr	0.60	9.30	26.38	5.35	41.03
100mm half round aluminium gutter to softwood fascia	32	m	5.40	83.70	239.94	48.55	372.19
extra over for							
running outlet	2	nr	0.30	4.65	15.92	3.09	23.66
Total			15.70	243.35	462.76	105.92	812.03

	Qty	Unit	Labour hours	Hours cost £	Materials £	O & P £	Total £
Two-storey terraced house size 7 x 7m with gable ends							
Take down existing rainwater goods and make good		Item	5.00	77.50	0.00	11.63	89.13
76mm diameter aluminium rainwater pipe with pipe clips plugged to brickwork at 2m maximum centres	12	m	3.60	55.80	245.88	45.25	346.93
extra over for							
offset	2	nr	0.60	9.30	57.58	10.03	76.91
shoe	2	nr	0.60	9.30	26.38	5.35	41.03
100mm half round aluminium gutter to softwood fascia	14	m	3.20	49.60	186.62	35.43	271.65
extra over for							
running outlet	2	nr	0.60	9.30	23.78	4.96	38.04
Total			13.60	210.80	540.24	112.66	863.70

	Qty	Unit	Labour hours	Hours cost £	Materials £	O & P £	Total £
Two-storey semi-detached house size 8 x 7m with gable ends							
Take down existing rainwater goods and make good		Item	7.00	108.50	0.00	16.28	124.78
76mm diameter aluminium rainwater pipe with pipe clips plugged to brickwork at 2m maximum centres	12	m	3.60	55.80	245.88	45.25	346.93
Carried forward			10.60	164.30	245.88	61.53	471.71

	Qty	Unit	Labour hours	Hours cost £	Materials £	O & P £	Total £
Brought forward			10.60	164.30	245.88	61.53	471.71
extra over for							
offset	2	nr	0.60	9.30	57.58	10.03	76.91
shoe	2	nr	0.60	9.30	26.38	5.35	41.03
100mm half round aluminium gutter to softwood fascia	16	m	4.80	74.40	213.38	43.17	330.95
extra over for							
stop end outlet	2	nr	0.30	4.65	15.72	3.06	23.43
Total			16.90	261.95	558.94	123.13	944.02

	Qty	Unit	Labour hours	Hours cost £	Materials £	O & P £	Total £
Two-storey detached house size 9 x 8m with gable ends							
Take down existing rainwater goods and make good		Item	8.00	124.00	0.00	18.60	142.60
76mm diameter aluminium rainwater pipe with pipe clips plugged to brickwork at 2m maximum centres	12	m	3.60	55.80	245.88	45.25	346.93
extra over for							
offset	2	nr	0.60	9.30	57.58	10.03	76.91
shoe	2	nr	0.60	9.30	26.38	5.35	41.03
100mm half round aluminium gutter to softwood fascia	18	m	5.40	83.70	239.94	48.55	372.19
extra over for							
stop end outlet	2	nr	0.30	4.65	15.72	3.06	23.43
Total			18.50	286.75	585.50	130.84	1,003.09

	Qty	Unit	Labour hours	Hours cost £	Materials £	O & P £	Total £
Three-storey terraced house size 8 x 8m with gable ends							
Take down existing rainwater goods and make good		Item	6.00	93.00	0.00	13.95	106.95
76mm diameter aluminium rainwater pipe with pipe clips plugged to brickwork at 2m maximum centres	18	m	5.40	83.70	368.72	67.86	520.28
extra over for							
offset	2	nr	0.60	9.30	57.58	10.03	76.91
shoe	2	nr	0.60	9.30	26.38	5.35	41.03
100mm half round aluminium gutter to softwood fascia	16	m	4.80	74.40	213.28	43.15	330.83
extra over for							
running outlet	2	nr	0.60	9.30	23.78	4.96	38.04
Total			18.00	279.00	689.74	145.31	1,114.05

	Qty	Unit	Labour hours	Hours cost £	Materials £	O & P £	Total £
Three-storey semi-detached house size 9 x 8m with gable ends							
Take down existing rainwater goods and make good		Item	8.00	124.00	0.00	18.60	142.60
75mm diameter aluminium rainwater pipe with pipe clips plugged to brickwork at 2m maximum centres	18	m	5.40	83.70	368.82	67.88	520.40
Carried forward			13.40	207.70	368.82	86.48	663.00

	Qty	Unit	Labour hours	Hours cost £	Materials £	O & P £	Total £
Brought forward			13.40	207.70	368.82	86.48	663.00
extra over for							
offset	2	nr	0.60	9.30	57.58	10.03	76.91
shoe	2	nr	0.60	9.30	26.38	5.35	41.03
100mm half round aluminium gutter to softwood fascia	18	m	5.40	83.70	239.94	48.55	372.19
extra over for							
stop end outlet	2	nr	0.30	4.65	15.72	3.06	23.43
Total			20.30	314.65	708.44	153.46	1,176.55

	Qty	Unit	Labour hours	Hours cost £	Materials £	O & P £	Total £
Three-storey detached house size 10 x 9m with gable ends							
Take down existing rainwater goods and make good		Item	9.00	139.50	0.00	20.93	160.43
75mm diameter aluminium rainwater pipe with pipe clips plugged to brickwork at 2m maximum centres	18	m	5.40	83.70	368.82	67.88	520.40
extra over for							
offset	2	nr	0.60	9.30	57.58	10.03	76.91
shoe	2	nr	0.60	9.30	26.38	5.35	41.03
100mm half round aluminium gutter to softwood fascia	20	m	6.00	93.00	266.60	53.94	413.54
extra over for							
stop end outlet	2	nr	0.30	4.65	15.72	3.06	23.43
Total			21.90	339.45	735.10	161.18	1,235.73

	Qty	Unit	Labour hours	Hours cost £	Materials £	O & P £	Total £
One-storey terraced house size 7 x 6m with hipped ends							
Take down existing rainwater goods and make good		Item	5.00	77.50	0.00	11.63	89.13
75mm diameter aluminium rainwater pipe with pipe clips plugged to brickwork at 2m maximum centres	6	m	1.80	27.90	122.94	22.63	173.47
extra over for							
offset	2	nr	0.60	9.30	57.58	10.03	76.91
shoe	2	nr	0.60	9.30	26.38	5.35	41.03
100mm half round aluminium gutter to softwood fascia	14	m	3.20	49.60	186.62	35.43	271.65
extra over for							
running outlet	2	nr	0.60	9.30	23.78	4.96	38.04
Total			11.80	182.90	417.30	90.03	690.23
One-storey semi-detached house size 8 x 7m with hipped ends							
Take down existing rainwater goods and make good		Item	6.00	93.00	0.00	13.95	106.95
75mm diameter aluminium rainwater pipe with pipe clips plugged to brickwork at 2m maximum centres	6	m	1.80	27.90	122.94	22.63	173.47
extra over for							
offset	2	nr	0.60	8.70	57.58	9.94	76.22
shoe	2	nr	0.60	9.30	26.38	5.35	41.03
Carried forward			9.00	138.90	206.90	51.87	397.67

	Qty	Unit	Labour hours	Hours cost £	Materials £	O & P £	Total £
Brought forward			9.00	138.90	206.90	51.87	397.67
100mm half round aluminium gutter to softwood fascia	23	m	6.90	106.95	306.59	62.03	475.57
extra over for							
angle	2	nr	0.60	9.30	13.98	3.49	26.77
running outlet	2	nr	0.60	9.30	23.78	4.96	38.04
Total			17.10	264.45	551.25	122.36	938.06

	Qty	Unit	Labour hours	Hours cost £	Materials £	O & P £	Total £
One-storey detached house size 9 x 7m with hipped ends							
Take down existing rainwater goods and make good		Item	8.00	124.00	0.00	18.60	142.60
75mm diameter aluminium rainwater pipe with pipe clips plugged to brickwork at 2m maximum centres	6	m	1.80	27.90	122.94	22.63	173.47
extra over for							
offset	2	nr	0.60	9.30	57.58	10.03	76.91
shoe	2	nr	0.60	9.30	26.38	5.35	41.03
100mm half round aluminium gutter to softwood fascia	32	m	9.60	148.80	426.56	86.30	661.66
extra over for							
angle	4	nr	0.60	9.30	27.96	5.59	42.85
running outlet	2	nr	1.20	18.60	23.78	6.36	48.74
Total			22.40	347.20	685.20	154.86	1,187.26

	Qty	Unit	Labour hours	Hours cost £	Materials £	O & P £	Total £
Two-storey terraced house size 7 x 7m with hipped ends							
Take down existing rainwater goods and make good		Item	6.00	93.00	0.00	13.95	106.95
75mm diameter aluminium rainwater pipe with pipe clips plugged to brickwork at 2m maximum centres	12	m	3.60	55.80	245.88	45.25	346.93
extra over for							
offset	2	nr	0.60	9.30	57.58	10.03	76.91
shoe	2	nr	0.60	9.30	26.38	5.35	41.03
100mm half round aluminium gutter to softwood fascia	14	m	3.20	49.60	186.62	35.43	271.65
extra over for							
running outlet	2	nr	0.60	9.30	23.78	4.96	38.04
Total			14.60	226.30	540.24	114.98	881.52

	Qty	Unit	Labour hours	Hours cost £	Materials £	O & P £	Total £
Two-storey semi-detached house size 8 x 7m with hipped ends							
Take down existing rainwater goods and make good		Item	8.00	124.00	0.00	18.60	142.60
75mm diameter aluminium rainwater pipe with pipe clips plugged to brickwork at 2m maximum centres	12	m	3.60	55.80	245.88	45.25	346.93
Carried forward			11.60	179.80	245.88	63.85	489.53

	Qty	Unit	Labour hours	Hours cost £	Materials £	O & P £	Total £
Brought forward			11.60	179.80	245.88	63.85	489.53
extra over for							
offset	2	nr	0.60	9.30	57.58	10.03	76.91
shoe	2	nr	0.60	9.30	26.38	5.35	41.03
100mm half round aluminium gutter to softwood fascia	23	m	6.90	106.95	306.29	61.99	475.23
extra over for							
angle	2	nr	0.60	9.30	23.78	4.96	38.04
running outlet	2	nr	0.60	9.30	13.98	3.49	26.77
Total			20.90	323.95	673.89	149.68	1,147.52

	Qty	Unit	Labour hours	Hours cost £	Materials £	O & P £	Total £
Two-storey detached house size 9 x 8m with hipped ends							
Take down existing rainwater goods and make good		Item	9.00	139.50	0.00	20.93	160.43
75mm diameter aluminium rainwater pipe as before	12	m	3.60	55.80	245.88	45.25	346.93
extra over for							
offset	2	nr	0.60	9.30	57.58	10.03	76.91
shoe	2	nr	0.60	9.30	26.38	5.35	41.03
100mm half round aluminium gutter to softwood fascia	34	m	10.20	158.10	423.22	87.20	668.52
extra over for							
angle	4	nr	0.60	9.30	27.96	5.59	42.85
running outlet	2	nr	1.20	18.60	23.78	6.36	48.74
Total			25.80	399.90	804.80	180.71	1,385.41

	Qty	Unit	Labour hours	Hours cost £	Materials £	O & P £	Total £
Three-storey terraced house size 8 x 8m with hipped ends							
Take down existing rainwater goods and make good		Item	7.00	108.50	0.00	16.28	124.78
75mm diameter aluminium rainwater pipe with pipe clips plugged to brickwork at 2m maximum centres	18	m	5.40	83.70	368.82	67.88	520.40
extra over for							
offset	2	nr	0.60	9.30	57.58	10.03	76.91
shoe	2	nr	0.60	9.30	24.38	5.05	38.73
100mm half round aluminium gutter to softwood fascia	16	m	4.80	74.40	213.28	43.15	330.83
extra over for							
running outlet	2	nr	0.60	9.30	23.78	4.96	38.04
Total			19.00	294.50	687.84	147.35	1,129.69

	Qty	Unit	Labour hours	Hours cost £	Materials £	O & P £	Total £
Three-storey semi-detached house size 9 x 8m with hipped ends							
Take down existing rainwater goods and make good		Item	9.00	139.50	0.00	20.93	160.43
68mm diameter PVC-U rainwater pipe with pipe clips plugged to brickwork at 2m maximum centres	18	m	5.40	83.70	368.82	67.88	520.40
Carried forward			14.40	223.20	368.82	88.80	680.82

	Qty	Unit	Labour hours	Hours cost £	Materials £	O & P £	Total £
Brought forward			14.40	223.20	368.82	88.80	680.82
extra over for							
offset	2	nr	0.60	9.30	57.58	10.03	76.91
shoe	2	nr	0.60	9.30	26.38	5.35	41.03
100mm half round aluminium gutter to softwood fascia	26	m	7.80	120.90	346.58	70.12	537.60
extra over for							
angle	2	nr	0.60	9.30	13.98	3.49	26.77
running outlet	2	nr	0.60	9.30	23.78	4.96	38.04
Total			24.60	381.30	837.12	182.76	1,401.18

	Qty	Unit	Labour hours	Hours cost £	Materials £	O & P £	Total £
Three-storey detached house size 10 x 9m with hipped ends							
Take down existing rainwater goods and make good		Item	10.00	155.00	0.00	23.25	178.25
75mm diameter cast iron rainwater pipe as before	18	m	5.40	83.70	368.82	67.88	520.40
extra over for							
offset	2	nr	0.60	9.30	57.58	10.03	76.91
shoe	2	nr	0.60	9.30	26.38	5.35	41.03
100mm half round cast iron gutter to softwood fascia	38	m	11.40	176.70	506.54	102.49	785.73
extra over for							
angle	4	nr	0.60	9.30	27.96	5.59	42.85
running outlet	2	nr	1.20	18.60	23.78	6.36	48.74
Total			29.80	461.90	1011.06	220.94	1,693.90

SUMMARY OF RAINWATER GOODS PROJECT COSTS

	Labour hours	Hours cost £	Materials £	O & P £	Total £
PVC-U RAINWATER GOODS					
One-storey gable end					
Terraced	9.82	152.21	129.22	42.21	323.64
Semi-detached	11.86	183.83	138.18	48.30	370.31
Detached	13.26	205.53	147.14	52.90	405.57
Two-storey gable end					
Terraced	12.32	190.96	165.70	53.50	410.16
Semi-detached	14.86	230.33	174.66	60.75	465.74
Detached	15.86	237.83	183.62	63.22	484.67
Three-storey gable end					
Terraced	15.22	235.91	211.14	67.06	514.11
Semi-detached	16.36	253.58	220.10	71.05	544.73
Detached	18.76	286.78	229.06	77.38	593.22
One-storey hipped end					
Terraced	9.82	152.21	129.22	42.21	323.64
Semi-detached	14.14	219.17	179.18	59.75	458.10
Detached	18.46	278.13	219.60	74.66	572.39
Two-storey hipped end					
Terraced	13.32	206.46	165.70	55.82	427.98
Semi-detached	17.64	273.42	215.66	73.36	562.44
Detached	21.36	322.08	266.30	88.26	676.64
Three-storey hipped end					
Terraced	16.22	251.41	211.14	69.38	531.93
Semi-detached	20.74	321.47	265.58	88.06	675.11
Detached	24.14	374.17	319.34	104.03	797.54

	Labour hours	Hours cost £	Materials £	O & P £	Total £
CAST IRON RAINWATER GOODS					
One-storey gable end					
Terraced	11.80	182.90	409.86	88.91	681.67
Semi-detached	14.10	218.55	437.96	98.48	754.99
Detached	15.70	243.35	469.96	107.00	820.31
Two-storey gable end					
Terraced	14.60	226.30	543.36	115.45	885.11
Semi-detached	16.90	261.95	575.36	125.60	962.91
Detached	18.50	286.75	603.46	133.53	1,023.74
Three-storey gable end					
Terraced	15.00	232.50	686.86	137.90	1,057.26
Semi-detached	19.30	299.15	708.86	151.20	1,159.21
Detached	21.90	339.45	736.96	161.46	1,237.87
One-storey hipped end					
Terraced	11.80	182.90	409.90	88.92	681.72
Semi-detached	17.10	265.05	459.84	108.73	833.62
Detached	22.40	347.20	702.44	157.45	1,207.09
Two-storey hipped end					
Terraced	15.60	241.80	543.36	117.77	902.93
Semi-detached	20.90	323.95	689.65	152.04	1,165.64
Detached	25.80	399.90	864.04	189.59	1,453.53
Three-storey hipped end					
Terraced	19.00	294.50	705.00	149.93	1,149.43
Semi-detached	24.60	381.30	865.30	186.99	1,433.59
Detached	29.80	461.90	1053.74	227.35	1,742.99

	Labour hours	Hours cost £	Materials £	O & P £	Total £
ALUMINIUM RAINWATER GOODS					
One-storey gable end					
Terraced	10.80	167.40	417.30	87.71	672.41
Semi-detached	14.10	218.55	435.90	98.17	752.62
Detached	15.70	243.35	462.76	105.92	812.03
Two-storey gable end					
Terraced	13.60	210.80	540.24	112.66	863.70
Semi-detached	16.90	261.95	558.94	123.13	944.03
Detached	18.50	286.75	585.50	130.84	1,003.09
Three-storey gable end					
Terraced	18.00	279.00	689.74	145.31	1,114.05
Semi-detached	20.30	314.65	708.44	153.46	1,176.55
Detached	21.90	339.45	735.10	161.18	1,235.73
Terraced	11.80	182.90	417.30	90.03	690.23
Semi-detached	17.10	264.45	551.25	122.36	938.06
Detached	22.40	347.20	685.20	154.86	1,187.26
Two-storey hipped end					
Terraced	14.60	226.30	540.24	114.98	881.52
Semi-detached	20.90	323.95	673.90	149.68	1,147.53
Detached	25.80	399.90	804.80	180.71	1,385.41
Three-storey hipped end					
Terraced	19.00	294.50	687.84	147.35	1,129.69
Semi-detached	24.60	381.30	837.12	182.76	1,401.18
Detached	29.80	461.90	1011.06	220.94	1,693.90

	Qty	Unit	Labour hours	Hours cost £	Materials £	O & P £	Total £
BATHROOMS							
Bathroom BCD with 1 bath with shower handset (B), 1 WC (C) and 1 lavatory basin (D)							
Stripping out							
Take out sanitary fittings from existing bathroom, including supply pipes, overflows and wastes, remove debris and make good to floors and walls to receive new fittings and pipework		Item	6.00	93.00	15.00	16.20	124.20
Pipework							
15mm diameter copper supply pipe	11.30	m	2.26	35.03	20.52	8.33	63.88
extra over for							
elbow	9	nr	1.62	25.11	9.00	5.12	39.23
tee	3	nr	0.66	10.23	3.00	1.98	15.21
tap connector	4	nr	0.36	5.58	7.72	2.00	15.30
19mm diameter MPVC-U overflow pipe	0.50	m	0.10	1.55	0.93	0.37	2.85
extra over for							
elbow	1	nr	0.18	2.79	1.37	0.62	4.78
straight connector	1	nr	0.20	3.10	2.03	0.77	5.90
32mm diameter polypropylene waste pipe	1	m	0.20	3.10	3.29	0.96	7.35
Carried forward			11.58	179.49	62.86	36.35	278.70

	Qty	Unit	Labour hours	Hours cost £	Materials £	O & P £	Total £
Brought forward			11.58	179.49	62.86	36.35	278.70
extra over for bend	2	nr	0.36	5.58	4.00	1.44	11.02
Traps							
32mm diameter polypropylene P trap	2	nr	0.60	9.30	9.14	2.77	21.21
Sanitary fittings							
Acrylic reinforced bath 1700mm long complete with chromium-plated grip handles, 40mm waste fitting, overflow, chain, plug bath panels, 20mm chromium-plated taps and shower handset	1	nr	2.75	42.63	261.32	45.59	349.54
Vitreous china-low level WC suite comprising pan, plastic seat and cover, 9 litre cistern and brackets and plastic connecting pipe	1	nr	2.45	37.98	222.97	39.14	300.09
Vitreous china wash basin size 560 × 430mm, complete with 32mm waste fitting, overflow, chain, stay and plug, pair 13mm chromium-plated easy clean pillar taps, cast iron cantilever brackets and pedestal	1	nr	2.20	34.10	131.51	24.84	190.45
Total			19.94	309.07	691.80	150.13	1,151.00

	Qty	Unit	Labour hours	Hours cost £	Materials £	O & P £	Total £
Bathroom BCDF with 1 bath with shower handset (B), 1 WC (C), 1 lavatory basin (D) and 1 bidet (F)							
Stripping out							
Take out sanitary fittings from existing bathroom, including supply pipes, overflows and wastes, remove debris and make good to floors and walls to receive new fittings and pipework		Item	6.00	93.00	15.00	16.20	124.20
Pipework							
15mm diameter copper supply pipe	14.10	m	2.82	43.71	25.66	10.41	79.78
extra over for							
elbow	12	nr	2.16	33.48	12.00	6.82	52.30
tee	3	nr	0.66	10.23	3.00	1.98	15.21
tap connector	3	nr	0.54	8.37	11.58	2.99	22.94
19mm diameter MPVC-U overflow pipe	1	m	0.20	3.10	1.87	0.75	5.72
extra over for							
elbow	2	nr	0.36	5.58	2.74	1.25	9.57
straight connector	2	nr	0.40	6.20	4.06	1.54	11.80
32mm diameter polypropylene waste pipe	1	m	0.20	3.10	3.29	0.96	7.35
extra over for							
bend	2	nr	0.36	5.58	4.00	1.44	11.02
Carried forward			13.70	212.35	83.20	44.33	339.88

	Qty	Unit	Labour hours	Hours cost £	Materials £	O & P £	Total £
Brought forward			13.70	212.35	83.20	44.33	339.88
Traps							
32mm diameter polypropylene P trap	2	nr	0.60	9.30	9.14	2.77	21.21
Sanitary fittings							
Acrylic reinforced bath 1700mm long complete with chromium-plated grip handles, 40mm waste fitting, overflow, chain, plug bath panels, 20mm chromium-plated taps and shower handset	1	nr	2.75	42.63	261.32	45.59	349.54
Vitreous china-low level WC suite comprising pan, plastic seat and cover, 9 litre cistern and brackets and plastic connecting pipe	1	nr	2.45	37.98	222.97	39.14	300.09
Vitreous china wash basin size 560 × 430mm, complete with 32mm waste fitting, overflow, chain, stay and plug, pair 13mm chromium-plated easy clean pillar taps, cast iron cantilever brackets and pedestal	1	nr	2.20	34.10	131.51	24.84	190.45
Bidet, free-standing with plain rim and chromium-plated monobloc spray and waste fittings complete	1	nr	2.50	38.75	282.08	48.12	368.95
Total			24.20	375.10	990.22	204.80	1,570.12

	Qty	Unit	Labour hours	Hours cost £	Materials £	O & P £	Total £
Bathroom BDF with 1 bath with shower handset (B), 1 lavatory basin (D) and 1 bidet (F)							
Stripping out							
Take out sanitary fittings from existing bathroom, including supply pipes, overflows and wastes, remove debris and make good to floors and walls to receive new fittings and pipework		Item	6.00	93.00	15.00	16.20	124.20
Pipework							
15mm diameter copper supply pipe	13	m	2.60	40.30	23.66	9.59	73.55
extra over for							
elbow	10	nr	1.80	27.90	10.00	5.69	43.59
tee	4	nr	0.88	13.64	4.00	2.65	20.29
tap connector	6	nr	1.08	16.74	23.16	5.99	45.89
19mm diameter MPVC-U overflow pipe	1	m	0.20	3.10	1.87	0.75	5.72
extra over for							
elbow	2	nr	0.36	5.58	2.74	1.25	9.57
straight connector	2	nr	0.40	6.20	4.06	1.54	11.80
32mm diameter polypropylene waste pipe	1	m	0.20	3.10	3.29	0.96	7.35
extra over for							
bend	2	nr	0.36	5.58	4.00	1.44	11.02
Carried forward			13.88	215.14	91.78	46.04	352.96

	Qty	Unit	Labour hours	Hours cost £	Materials £	O & P £	Total £
Brought forward			13.88	215.14	91.78	46.04	352.96
Traps							
32mm diameter polypropylene P trap	2	nr	0.60	9.30	9.14	2.77	21.21
Sanitary fittings							
Acrylic reinforced bath 1700mm long complete with chromium-plated grip handles, 40mm waste fitting, overflow, chain, plug bath panels, 20mm chromium-plated taps and shower handset	1	nr	2.75	42.63	261.32	45.59	349.54
Vitreous china wash basin size 560 × 430mm, complete with 32mm waste fitting, overflow, chain, stay and plug, pair 13mm chromium-plated easy clean pillar taps, cast iron cantilever brackets and pedestal	1	nr	2.20	34.10	131.50	24.84	190.44
Bidet, free-standing with plain rim and chromium-plated monobloc spray and waste fittings complete	1	nr	2.50	38.75	258.79	44.63	342.17
Total			21.93	339.92	752.53	163.87	1,256.31

	Qty	Unit	Labour hours	Hours cost £	Materials £	O & P £	Total £
Bathroom BD with 1 bath with shower handset (B) and 1 lavatory basin (D)							
Stripping out							
Take out sanitary fittings from existing bathroom, including supply pipes, overflows and wastes, remove debris and make good to floors and walls to receive new fittings and pipework		Item	6.00	93.00	15.00	16.20	124.20
Pipework							
15mm diameter copper supply pipe	13.40	m	2.68	41.54	24.29	9.87	75.70
extra over for							
elbow	10	nr	1.80	27.90	10.00	5.69	43.59
tee	2	nr	0.44	6.82	2.00	1.32	10.14
tap connector	4	nr	0.72	11.16	15.44	3.99	30.59
19mm diameter MPVC-U overflow pipe	0.50	m	0.10	1.55	0.93	0.37	2.85
extra over for							
elbow	2	nr	0.18	2.79	1.37	0.62	4.78
straight connector	2	nr	0.20	3.10	2.03	0.77	5.90
32mm diameter polypropylene waste pipe	1	m	0.20	3.10	3.29	0.96	7.35
extra over for							
bend	2	nr	0.36	5.58	4.00	1.44	11.02
Carried forward			12.68	196.54	78.35	41.23	316.12

	Qty	Unit	Labour hours	Hours cost £	Materials £	O & P £	Total £
Brought forward			12.68	196.54	78.35	41.23	316.12
Traps							
32mm diameter polypropylene P trap	2	nr	0.60	9.30	9.14	2.77	21.21
Sanitary fittings							
Acrylic reinforced bath 1700mm long complete with chromium-plated grip handles, 40mm waste fitting, overflow, chain, plug bath panels, 20mm chromium-plated taps and shower handset	1	nr	2.75	42.63	261.32	45.59	349.54
Vitreous china wash basin size 560 × 430mm, complete with 32mm waste fitting, overflow, chain, stay and plug, pair 13mm chromium-plated easy clean pillar taps, cast iron cantilever brackets and pedestal	1	nr	2.20	34.10	131.51	24.84	190.45
Total			18.23	282.57	480.32	114.43	877.31

	Qty	Unit	Labour hours	Hours cost £	Materials £	O & P £	Total £
Bathroom BCDE with 1 bath with shower handset (B), 1 WC (C), 1 lavatory basin (D) and 1 shower cubicle (E)							
Stripping out							
Take out sanitary fittings from existing bathroom, including supply pipes, overflows and wastes, remove debris and make good to floors and walls to receive new fittings and pipework		Item	6.00	93.00	15.00	16.20	124.20
Pipework							
15mm diameter copper supply pipe	16.30	m	3.26	50.53	29.37	11.99	91.89
extra over for							
elbow	11	nr	1.98	30.69	11.00	6.25	47.94
tee	5	nr	1.10	17.05	5.00	3.31	25.36
tap connector	4	nr	0.72	11.16	15.44	3.99	30.59
19mm diameter MPVC-U overflow pipe	0.50	m	0.50	7.75	0.93	1.30	9.98
extra over for							
elbow	1	nr	0.18	2.79	1.37	0.62	4.78
straight connector	1	nr	0.20	3.10	2.03	0.77	5.90
32mm diameter polypropylene waste pipe	1.50	m	0.30	4.65	4.94	1.44	11.03
extra over for							
bend	2	nr	0.36	5.58	4.00	1.44	11.02
Carried forward			14.60	226.30	89.08	47.31	362.69

	Qty	Unit	Labour hours	Hours cost £	Materials £	O & P £	Total £
Brought forward			14.60	226.30	89.08	47.31	362.69
Traps							
32mm diameter polypropylene P trap	3	nr	0.90	13.95	13.71	4.15	31.81
Sanitary fittings							
Acrylic reinforced bath 1700mm long complete with chromium-plated grip handles, 40mm waste fitting, overflow, chain, plug bath panels, 20mm chromium-plated taps and shower handset	1	nr	2.75	42.63	261.32	45.59	349.54
Vitreous china-low level WC suite comprising pan, plastic seat and cover, 9 litre cistern and brackets and plastic connecting pipe	1	nr	2.45	37.98	222.97	39.14	300.09
Vitreous china wash basin size 560 × 430mm, complete with 32mm waste fitting, overflow, chain, stay and plug, pair 13mm chromium-plated easy clean pillar taps, cast iron cantilever brackets and pedestal	1	nr	2.20	34.10	131.51	24.84	190.45
Carried forward			22.90	354.95	718.59	161.03	1,234.57

	Qty	Unit	Labour hours	Hours cost £	Materials £	O & P £	Total £
Brought forward			22.90	354.95	718.59	161.03	1,234.57
Shower cubicle size 788 × 842 × 2115mm high, anodised aluminium frame and safety glass, complete with 8.15kW instant electric shower with adjustable spray handset, slide bar and hose, soap dish and chromium-plated waste fittings	1	nr	3.80	58.90	751.53	121.56	931.99
Total			26.70	413.85	1,470.12	282.59	2,166.56

	Qty	Unit	Labour hours	Hours cost £	Materials £	O & P £	Total £
Bathroom BCDEF with 1 bath with shower handset (B), 1 WC (C), 1 lavatory basin (D), 1 shower cubicle (E) and 1 bidet (F)							
Stripping out							
Take out sanitary fittings from existing bathroom, including supply pipes, overflows and wastes, remove debris and make good to floors and walls to receive new fittings and pipework		Item	6.00	93.00	15.00	16.20	124.20
Pipework							
15mm diameter copper supply pipe	19.90	m	3.98	61.69	34.22	14.39	110.30
Carried forward			9.98	154.69	49.22	30.59	234.50

	Qty	Unit	Labour hours	Hours cost £	Materials £	O & P £	Total £
Brought forward			9.98	154.69	49.22	30.59	234.50
extra over for							
elbow	14	nr	2.52	39.06	14.00	7.96	61.02
tee	7	nr	1.54	23.87	7.00	4.63	35.50
tap connector	6	nr	0.72	11.16	23.16	5.15	39.47
19mm diameter MPVC-U overflow pipe	1	m	0.20	3.10	1.87	0.75	5.72
extra over for							
elbow	2	nr	0.36	5.58	2.74	1.25	9.57
straight connector	2	nr	0.40	6.20	2.03	1.23	9.46
32mm diameter polypropylene waste pipe	1.50	m	0.30	4.65	4.94	1.44	11.03
extra over for							
bend	2	nr	0.36	5.58	4.00	1.44	11.02
Traps							
32mm diameter polypropylene P trap	3	nr	0.90	13.95	13.71	4.15	31.81
Sanitary fittings							
Acrylic reinforced bath 1700mm long complete with chromium-plated grip handles, 40mm waste fitting, overflow, chain, plug bath panels, 20mm chromium-plated taps and shower handset	1	nr	2.75	42.63	261.32	45.59	349.54
Carried forward			20.03	310.47	383.99	104.17	798.63

	Qty	Unit	Labour hours	Hours cost £	Materials £	O & P £	Total £
Brought forward			20.03	310.47	383.99	104.17	798.63
Vitreous china-low level WC suite comprising pan, plastic seat and cover, 9 litre cistern and brackets and plastic connecting pipe	1	nr	2.45	37.98	222.97	39.14	300.09
Vitreous china wash basin size 560 × 430mm, complete with 32mm waste fitting, overflow, chain, stay and plug, pair 13mm chromium-plated easy clean pillar taps, cast iron cantilever brackets and pedestal	1	nr	2.20	34.10	131.51	24.84	190.45
Shower cubicle size 788 × 842 × 2115mm high, anodised aluminium frame and safety glass, complete with 8.15kW instant electric shower with adjustable spray handset, slide bar and hose, soap dish and chromium-plated waste fittings	1	nr	3.80	58.90	751.53	121.56	931.99
Bidet, free-standing with plain rim and chromium-plated monobloc spray and waste fittings complete	1	nr	2.50	38.75	282.08	48.12	368.95
Total			30.98	480.20	1,772.08	337.84	2,590.12

	Qty	Unit	Labour hours	Hours cost £	Materials £	O & P £	Total £
Bathroom BDEF with 1 bath with shower handset (B), 1 lavatory basin (D), 1 shower cubicle (E) and 1 bidet (F)							
Stripping out							
Take out sanitary fittings from existing bathroom, including supply pipes, overflows and wastes, remove debris and make good to floors and walls to receive new fittings and pipework		Item	6.00	93.00	15.00	16.20	124.20
Pipework							
15mm diameter copper supply pipe	15.80	m	3.16	48.98	28.76	11.66	89.40
extra over for							
elbow	12	nr	2.16	33.48	12.00	6.82	52.30
tee	6	nr	1.32	20.46	6.00	3.97	30.43
tap connector	6	nr	1.08	16.74	23.16	5.99	45.89
19mm diameter MPVC-U overflow pipe	1	m	0.20	3.10	1.87	0.75	5.72
extra over for							
elbow	2	nr	0.36	5.58	2.74	1.25	9.57
straight connector	2	nr	0.40	6.20	2.03	1.23	9.46
32mm diameter polypropylene waste pipe	1.50	m	0.30	4.65	4.94	1.44	11.03
extra over for							
bend	2	nr	0.36	5.58	4.00	1.44	11.02
Carried forward			15.34	237.77	100.50	50.74	389.01

	Qty	Unit	Labour hours	Hours cost £	Materials £	O & P £	Total £
Brought forward			15.34	237.77	100.50	50.74	389.01
Traps							
32mm diameter polypropylene P trap	3	nr	0.90	13.95	13.71	4.15	31.81
Sanitary fittings							
Acrylic reinforced bath 1700mm long complete with chromium-plated grip handles, 40mm waste fitting, overflow, chain, plug bath panels, 20mm chromium-plated taps and shower handset	1	nr	2.75	42.63	261.32	45.59	349.54
Vitreous china wash basin size 560 × 430mm, complete with 32mm waste fitting, overflow, chain, stay and plug, pair 13mm chromium-plated easy clean pillar taps, cast iron cantilever brackets and pedestal	1	nr	2.20	34.10	222.97	38.56	295.63
Shower cubicle size 788 × 842 × 2115mm high, anodised aluminium frame and safety glass, complete with 8.15kW instant electric shower with adjustable spray handset, slide bar and hose, soap dish and chromium-plated waste fittings	1	nr	3.80	58.90	751.53	121.56	931.99
Carried forward			24.99	387.35	1,350.03	260.61	1,997.98

	Qty	Unit	Labour hours	Hours cost £	Materials £	O & P £	Total £
Brought forward			24.99	387.35	1,350.03	260.61	1,997.98
Bidet, free-standing with plain rim and chromium-plated monobloc spray and waste fittings complete	1	nr	2.50	38.75	282.08	48.12	368.95
Total			27.49	426.10	1,632.11	308.73	2,366.93

	Qty	Unit	Labour hours	Hours cost £	Materials £	O & P £	Total £
Bathroom BDE with 1 bath with shower handset (B), 1 lavatory basin (D) and 1 shower cubicle (E)							
Stripping out							
Take out sanitary fittings from existing bathroom, including supply pipes, overflows and wastes, remove debris and make good to floors and walls to receive new fittings and pipework		Item	6.00	93.00	15.00	16.20	124.20
Pipework							
15mm diameter copper supply pipe	15.80	m	3.16	48.98	28.76	11.66	89.40
extra over for							
elbow	10	nr	1.80	27.90	10.00	5.69	43.59
tee	4	nr	0.88	13.64	4.00	2.65	20.29
tap connector	6	nr	1.08	16.74	23.16	5.99	45.89
Carried forward			12.92	200.26	80.92	42.18	323.36

	Qty	Unit	Labour hours	Hours cost £	Materials £	O & P £	Total £
Brought forward			12.92	200.26	80.92	42.18	323.36
19mm diameter MPVC-U overflow pipe	0.50	m	0.10	1.55	0.93	0.37	2.85
extra over for							
elbow	1	nr	0.18	2.79	1.37	0.62	4.78
straight connector	1	nr	0.20	3.10	2.03	0.77	5.90
32mm diameter polypropylene waste pipe	1.00	m	0.20	3.10	3.29	0.96	7.35
extra over for							
bend	2	nr	0.36	5.58	4.00	1.44	11.02
Traps							
32mm diameter polypropylene P trap	3	nr	0.90	13.95	13.71	4.15	31.81
Sanitary fittings							
Acrylic reinforced bath 1700mm long complete with chromium-plated grip handles, 40mm waste fitting, overflow, chain, plug bath panels, 20mm chromium-plated taps and shower handset	1	nr	2.75	42.63	261.32	45.59	349.54
Vitreous china wash basin size 560 × 430mm, complete with 32mm waste fitting, overflow, chain, stay and plug, pair 13mm chromium-plated easy clean pillar taps, cast iron cantilever brackets and pedestal	1	nr	2.20	34.10	131.51	24.84	190.45
Carried forward			19.81	307.06	499.08	120.92	927.06

	Qty	Unit	Labour hours	Hours cost £	Materials £	O & P £	Total £
Brought forward			19.81	307.06	499.08	120.92	927.06
Shower cubicle size 788 × 842 × 2115mm high, anodised aluminium frame and safety glass, complete with 8.15kW instant electric shower with adjustable spray handset, slide bar and hose, soap dish and chromium-plated waste fittings	1	nr	3.80	58.90	751.53	121.56	931.99
Total			23.61	365.96	1,250.61	242.48	1,859.05

	Qty	Unit	Labour hours	Hours cost £	Materials £	O & P £	Total £
Bathroom ACD with 1 bath (A), 1 WC (C) and 1 lavatory basin (D)							
Stripping out							
Take out sanitary fittings from existing bathroom, including supply pipes, overflows and wastes, remove debris and make good to floors and walls to receive new fittings and pipework		Item	6.00	93.00	15.00	16.20	124.20
Pipework							
15mm diameter copper supply pipe	11.30	m	2.26	35.03	20.52	8.33	63.88
Carried forward			8.26	128.03	35.52	24.53	188.08

	Qty	Unit	Labour hours	Hours cost £	Materials £	O & P £	Total £
Brought forward			8.26	128.03	35.52	24.53	188.08
extra over for							
elbow	9	nr	1.80	27.90	10.00	5.69	43.59
tee	3	nr	0.88	13.64	4.00	2.65	20.29
tap connector	4	nr	0.72	11.16	15.44	3.99	30.59
19mm diameter MPVC-U overflow pipe	0.50	m	0.10	1.55	1.87	0.51	3.93
extra over for							
elbow	1	nr	0.18	2.79	1.37	0.62	4.78
straight connector	1	nr	0.20	3.10	2.03	0.77	5.90
32mm diameter polypropylene waste pipe	1	m	0.20	3.10	3.29	0.96	7.35
extra over for							
bend	1	nr	0.36	5.58	4.00	1.44	11.02
Traps							
32mm diameter polypropylene P trap	1	nr	0.60	9.30	9.14	2.77	21.21
Sanitary fittings							
Acrylic reinforced bath 1700mm long complete with chromium-plated grip handles, 40mm waste fitting, overflow, chain, plug bath panels, 20mm chromium-plated taps	1	nr	2.75	42.63	186.65	34.39	263.67
Carried forward			16.05	248.78	273.31	78.31	600.40

	Qty	Unit	Labour hours	Hours cost £	Materials £	O & P £	Total £
Brought forward			16.05	248.78	273.31	78.31	600.40
Vitreous china-low level WC suite comprising pan, plastic seat and cover, 9 litre cistern and brackets and plastic connecting pipe	1	nr	2.45	37.98	222.97	39.14	300.09
Vitreous china wash basin size 560 × 430mm, complete with 32mm waste fitting, overflow, chain, stay and plug, pair 13mm chromium-plated easy clean pillar taps, cast iron cantilever brackets and pedestal	1	nr	2.20	34.10	131.51	24.84	190.45
Total			20.70	320.86	627.79	142.29	1,090.94

	Qty	Unit	Labour hours	Hours cost £	Materials £	O & P £	Total £
Bathroom ACDF with 1 bath (A), 1 WC (C), 1 lavatory basin (D) and 1 bidet (F)							
Stripping out							
Take out sanitary fittings from existing bathroom, including supply pipes, overflows and wastes, remove debris and make good to floors and walls to receive new fittings and pipework		Item	6.00	93.00	15.00	16.20	124.20
Carried forward			6.00	93.00	15.00	16.20	124.20

	Qty	Unit	Labour hours	Hours cost £	Materials £	O & P £	Total £
Brought forward			6.00	93.00	15.00	16.20	124.20
Pipework							
15mm diameter copper supply pipe	14.10	m	2.82	43.71	25.66	10.41	79.78
extra over for							
elbow	12	nr	2.16	33.48	12.00	6.82	52.30
tee	4	nr	0.66	10.23	3.00	1.98	15.21
tap connector	3	nr	0.54	8.37	11.58	2.99	22.94
19mm diameter MPVC-U overflow pipe	1	m	0.20	3.10	1.87	0.75	5.72
extra over for							
elbow	2	nr	0.36	5.58	2.74	1.25	9.57
straight connector	2	nr	0.40	6.20	4.06	1.54	11.80
32mm diameter polypropylene waste pipe	1	m	0.20	3.10	3.29	0.96	7.35
extra over for							
bend	2	nr	0.36	5.58	4.00	1.44	11.02
Traps							
32mm diameter polypropylene P trap	2	nr	0.60	9.30	9.14	2.77	21.21
Sanitary fittings							
Acrylic reinforced bath 1700mm long complete with chromium-plated grip handles, 40mm waste fitting, overflow, chain, plug bath panels, 20mm chromium-plated taps	1	nr	2.75	42.63	186.65	34.39	263.67
Carried forward			17.05	264.28	278.99	81.49	624.75

	Qty	Unit	Labour hours	Hours cost £	Materials £	O & P £	Total £
Brought forward			17.05	264.28	278.99	81.49	624.75
Vitreous china-low level WC suite comprising pan, plastic seat and cover, 9 litre cistern and brackets and plastic connecting pipe	1	nr	2.45	37.98	222.97	39.14	300.09
Vitreous china wash basin size 560 × 430mm, complete with 32mm waste fitting, overflow, chain, stay and plug, pair 13mm chromium-plated easy clean pillar taps, cast iron cantilever brackets and pedestal	1	nr	2.20	34.10	131.51	24.84	190.45
Bidet, free-standing with plain rim and chromium-plated monobloc spray and waste fittings complete	1	nr	2.50	38.75	282.08	48.12	368.95
Total			24.20	375.11	915.55	193.60	1,484.24

	Qty	Unit	Labour hours	Hours cost £	Materials £	O & P £	Total £
Bathroom ADF with 1 bath (A), 1 lavatory basin (D) and 1 bidet (F)							
Stripping out							
Take out sanitary fittings from existing bathroom, including supply pipes, overflows and wastes, remove debris and make good to floors and walls to receive new fittings and pipework		Item	6.00	93.00	15.00	16.20	124.20
Pipework							
15mm diameter copper supply pipe	13	m	2.60	40.30	23.26	9.53	73.09
extra over for							
elbow	10	nr	1.80	27.90	10.00	5.69	43.59
tee	4	nr	0.88	13.64	4.00	2.65	20.29
tap connector	6	nr	1.08	16.74	23.16	5.99	45.89
19mm diameter MPVC-U overflow pipe	1	m	0.25	3.88	1.87	0.86	6.61
extra over for							
elbow	2	nr	0.20	3.10	2.74	0.88	6.72
straight connector	2	nr	0.40	6.20	4.06	1.54	11.80
32mm diameter polypropylene waste pipe	1	m	0.36	5.58	3.29	1.33	10.20
extra over for							
bend	2	nr	0.40	6.20	4.00	1.53	11.73
Carried forward			13.97	216.54	91.38	46.19	354.10

	Qty	Unit	Labour hours	Hours cost £	Materials £	O & P £	Total £
Brought forward			13.97	216.54	91.38	46.19	354.10
Traps							
32mm diameter polypropylene P trap	2	nr	0.60	9.30	9.14	2.77	21.21
Sanitary fittings							
Acrylic reinforced bath 1700mm long complete with chromium-plated grip handles, 40mm waste fitting, overflow, chain, plug bath panels, 20mm chromium-plated taps	1	nr	2.75	42.63	186.65	34.39	263.67
Vitreous china wash basin size 560 × 430mm, complete with 32mm waste fitting, overflow, chain, stay and plug, pair 13mm chromium-plated easy clean pillar taps, cast iron cantilever brackets and pedestal	1	nr	2.45	37.98	131.51.	5.70	43.67
Vitreous china-low level WC suite comprising pan, plastic seat and cover, 9 litre cistern and brackets	1	nr	2.45	37.98	222.97	39.14	300.09
Total			22.22	344.42	510.14	128.19	982.73

	Qty	Unit	Labour hours	Hours cost £	Materials £	O & P £	Total £
Bathroom AD with 1 bath (A) and 1 lavatory basin (D)							
Stripping out							
Take out sanitary fittings from existing bathroom, including supply pipes, overflows and wastes, remove debris and make good to floors and walls to receive new fittings and pipework		Item	6.00	93.00	15.00	16.20	124.20
Pipework							
15mm diameter copper supply pipe	13.40	m	2.68	41.54	23.66	9.78	74.98
extra over for							
elbow	10	nr	1.80	27.90	10.00	5.69	43.59
tee	2	nr	0.44	6.82	4.00	1.62	12.44
tap connector	4	nr	0.72	11.16	23.16	5.15	39.47
19mm diameter MPVC-U overflow pipe	0.50	m	0.10	1.55	1.87	0.51	3.93
extra over for							
elbow	2	nr	0.18	2.79	2.74	0.83	6.36
straight connector	2	nr	0.20	3.10	4.06	1.07	8.23
32mm diameter polypropylene waste pipe	1	m	0.20	3.10	3.29	0.96	7.35
extra over for							
bend	2	nr	0.36	5.58	4.00	1.44	11.02
Carried forward			12.68	196.54	91.78	43.25	331.57

	Qty	Unit	Labour hours	Hours cost £	Materials £	O & P £	Total £
Brought forward			12.68	196.54	91.78	43.25	331.57
Traps							
32mm diameter polypropylene P trap	2	nr	0.60	9.30	9.14	2.77	21.21
Sanitary fittings							
Acrylic reinforced bath 1700mm long complete with chromium-plated grip handles, 40mm waste fitting, overflow, chain, plug bath panels, 20mm chromium-plated taps	1	nr	2.75	42.63	186.65	34.39	263.67
Vitreous china wash basin size 560 × 430mm, complete with 32mm waste fitting, overflow, chain, stay and plug, pair 13mm chromium-plated easy clean pillar taps, cast iron cantilever brackets and pedestal	1	nr	2.20	34.10	131.51	24.84	190.45
Total			18.23	282.57	419.08	105.25	806.89

	Qty	Unit	Labour hours	Hours cost £	Materials £	O & P £	Total £
Bathroom ACDE with 1 bath (A), 1 WC (C), 1 lavatory basin (D) and 1 shower cubicle (E)							
Stripping out							
Take out sanitary fittings from existing bathroom, including supply pipes, overflows and wastes, remove debris and make good to floors and walls to receive new fittings and pipework		Item	6.00	93.00	15.00	16.20	124.20
Pipework							
15mm diameter copper supply pipe	16.30	m	3.26	50.53	29.07	11.94	91.54
extra over for							
elbow	11	nr	1.98	30.69	11.00	6.25	47.94
tee	5	nr	1.10	17.05	5.00	3.31	25.36
tap connector	4	nr	0.72	11.16	15.44	3.99	30.59
19mm diameter MPVC-U overflow pipe	0.50	m	0.10	1.55	0.93	0.37	2.85
extra over for							
elbow	1	nr	0.18	2.79	1.37	0.62	4.78
straight connector	1	nr	0.20	3.10	2.03	0.77	5.90
32mm diameter polypropylene waste pipe	1.50	m	0.30	4.65	4.94	1.44	11.03
extra over for							
bend	2	nr	0.36	5.58	4.00	1.44	11.02
Carried forward			14.20	214.52	84.78	44.90	344.20

	Qty	Unit	Labour hours	Hours cost £	Materials £	O & P £	Total £
Brought forward			14.20	214.52	84.78	44.90	344.20
Traps							
32mm diameter polypropylene P trap	3	nr	0.90	13.95	9.14	3.46	26.55
Sanitary fittings							
Acrylic reinforced bath 1700mm long complete with chromium-plated grip handles, 40mm waste fitting, overflow, chain, plug bath panels, 20mm chromium-plated taps	1	nr	2.75	42.63	186.65	34.39	263.67
Vitreous china-low level WC suite comprising pan, plastic seat and cover, 9 litre cistern and brackets and plastic connecting pipe	1	nr	2.45	37.98	222.97	39.14	300.09
Vitreous china wash basin size 560 × 430mm, complete with 32mm waste fitting, overflow, chain, stay and plug, pair 13mm chromium-plated easy clean pillar taps, cast iron cantilever brackets and pedestal	1	nr	2.20	34.10	131.51	24.84	190.45
Carried forward			22.50	343.17	635.05	146.74	1,124.96

	Qty	Unit	Labour hours	Hours cost £	Materials £	O & P £	Total £
Brought forward			22.50	343.17	635.05	146.74	1,124.96
Shower cubicle size 788 × 842 × 2115mm high, anodised aluminium frame and safety glass, complete with 8.15kW instant electric shower with adjustable spray handset, slide bar and hose, soap dish and chromium-plated waste fittings	1	nr	3.80	58.90	751.53	121.56	931.99
Total			26.30	402.07	1,386.58	268.30	2,056.95

	Qty	Unit	Labour hours	Hours cost £	Materials £	O & P £	Total £
Bathroom ACDEF with 1 bath (A), 1 WC (C), 1 lavatory basin (D), 1 shower cubicle (E) and 1 bidet (F)							
Stripping out							
Take out sanitary fittings from existing bathroom, including supply pipes, overflows and wastes, remove debris and make good to floors and walls to receive new fittings and pipework		Item	6.00	93.00	15.00	16.20	124.20
Pipework							
15mm diameter copper supply pipe	19.90	m	3.98	61.69	36.04	14.66	112.39
Carried forward			9.98	154.69	51.04	30.86	236.59

	Qty	Unit	Labour hours	Hours cost £	Materials £	O & P £	Total £
Brought forward			9.98	154.69	51.04	30.86	236.59
extra over for							
elbow	14	nr	2.52	39.06	14.00	7.96	61.02
tee	5	nr	1.54	23.87	7.00	4.63	35.50
tap connector	4	nr	1.08	16.74	23.16	5.99	45.89
19mm diameter MPVC-U overflow pipe	1	m	0.20	3.10	1.87	0.75	5.72
extra over for							
elbow	1	nr	0.36	5.58	2.74	1.25	9.57
straight connector	1	nr	0.40	6.20	2.03	1.23	9.46
32mm diameter polypropylene waste pipe	1.50	m	0.30	4.65	4.94	1.44	11.03
extra over for							
bend	2	nr	0.36	5.58	4.00	1.44	11.02
Traps							
32mm diameter polypropylene P trap	3	nr	0.90	13.95	13.71	4.15	31.81
Sanitary fittings							
Acrylic reinforced bath 1700mm long complete with chromium-plated grip handles, 40mm waste fitting, overflow, chain, plug bath panels, 20mm chromium-plated taps	1	nr	2.75	42.63	186.65	34.39	263.67
Carried forward			20.39	316.05	311.14	94.08	721.26

	Qty	Unit	Labour hours	Hours cost £	Materials £	O & P £	Total £
Brought forward			20.39	316.05	311.14	94.08	721.26
Vitreous china-low level WC suite comprising pan, plastic seat and cover, 9 litre cistern and brackets and plastic connecting pipe	1	nr	2.45	37.98	222.97	39.14	300.09
Vitreous china wash basin size 560 × 430mm, complete with 32mm waste fitting, overflow, chain, stay and plug, pair 13mm chromium-plated easy clean pillar taps, cast iron cantilever brackets and pedestal	1	nr	2.20	34.10	131.51	24.84	190.45
Shower cubicle size 788 × 842 × 2115mm high, anodised aluminium frame and safety glass, complete with 8.15kW instant electric shower with adjustable spray handset, slide bar and hose, soap dish and chromium-plated waste fittings	1	nr	3.80	58.90	751.53	121.56	931.99
Bidet, free-standing with plain rim and chromium-plated monobloc spray and waste fittings complete	1	nr	2.50	38.75	282.08	48.12	368.95
Total			31.34	485.78	1,699.23	327.75	2,512.75

	Qty	Unit	Labour hours	Hours cost £	Materials £	O & P £	Total £
Bathroom ADEF with 1 bath (A), 1 lavatory basin (D), 1 shower cubicle (E) and 1 bidet (F)							
Stripping out							
Take out sanitary fittings from existing bathroom, including supply pipes, overflows and wastes, remove debris and make good to floors and walls to receive new fittings and pipework		Item	6.00	93.00	15.00	16.20	124.20
Pipework							
15mm diameter copper supply pipe	15.80	m	3.16	48.98	28.76	11.66	89.40
extra over for							
elbow	12	nr	2.16	33.48	12.00	6.82	52.30
tee	6	nr	1.32	20.46	6.00	3.97	30.43
tap connector	6	nr	1.08	16.74	23.16	5.99	45.89
19mm diameter MPVC-U overflow pipe	1	m	0.20	3.10	1.87	0.75	5.72
extra over for							
elbow	2	nr	0.36	5.58	2.74	1.25	9.57
straight connector	2	nr	0.40	6.20	2.03	1.23	9.46
32mm diameter polypropylene waste pipe	1.50	m	0.30	4.65	4.94	1.44	11.03
extra over for							
bend	2	nr	0.36	5.58	4.00	1.44	11.02
Carried forward			15.34	237.77	100.50	50.74	389.01

	Qty	Unit	Labour hours	Hours cost £	Materials £	O & P £	Total £
Brought forward			15.34	237.77	100.50	50.74	389.01
Traps							
32mm diameter polypropylene P trap	3	nr	0.90	13.95	13.71	4.15	31.81
Sanitary fittings							
Acrylic reinforced bath 1700mm long complete with chromium-plated grip handles, 40mm waste fitting, overflow, chain, plug bath panels, 20mm chromium-plated taps	1	nr	2.75	42.63	186.65	34.39	263.67
Vitreous china wash basin size 560 × 430mm, complete with 32mm waste fitting, overflow, chain, stay and plug, pair 13mm chromium-plated easy clean pillar taps, cast iron cantilever brackets and pedestal	1	nr	2.20	34.10	131.51	24.84	190.45
Shower cubicle size 788 × 842 × 2115mm high, anodised aluminium frame and safety glass, complete with 8.15kW instant electric shower with adjustable spray handset, slide bar and hose, soap dish and chromium-plated waste fittings	1	nr	3.80	58.90	751.53	121.56	931.99
Carried forward			24.99	387.35	1,183.90	235.69	1,806.93

	Qty	Unit	Labour hours	Hours cost £	Materials £	O & P £	Total £
Brought forward			24.99	387.35	1,183.90	235.69	1,806.93
Bidet, free-standing with plain rim and chromium plated monobloc spray and waste fittings complete	1	nr	2.50	38.75	282.08	48.12	368.95
Total			27.49	426.10	1,465.98	283.81	2,175.88

	Qty	Unit	Labour hours	Hours cost £	Materials £	O & P £	Total £
Bathroom ADE with 1 bath (A), 1 lavatory basin (D) and 1 shower cubicle (E)							
Stripping out							
Take out sanitary fittings from existing bathroom, including supply pipes, overflows and wastes, remove debris and make good to floors and walls to receive new fittings and pipework		Item	6.00	93.00	15.00	16.20	124.20
Pipework							
15mm diameter copper supply pipe	15.80	m	3.16	48.98	28.76	11.66	89.40
extra over for							
elbow	10	nr	1.80	27.90	10.00	5.69	43.59
tee	4	nr	0.88	13.64	4.00	2.65	20.29
tap connector	6	nr	1.08	16.74	23.16	5.99	45.89
Carried forward			12.92	200.26	80.92	42.18	323.36

	Qty	Unit	Labour hours	Hours cost £	Materials £	O & P £	Total £
Brought forward			12.92	200.26	80.92	42.18	323.36
19mm diameter MPVC-U overflow pipe	0.50	m	0.10	1.55	0.93	0.37	2.85
extra over for							
elbow	1	nr	0.18	2.79	1.37	0.62	4.78
straight connector	1	nr	0.20	3.10	2.03	0.77	5.90
32mm diameter polypropylene waste pipe	1.00	m	0.20	3.10	3.29	0.96	7.35
extra over for							
bend	2	nr	0.36	5.58	4.00	1.44	11.02
Traps							
32mm diameter polypropylene P trap	3	nr	0.90	13.95	13.71	4.15	31.81
Sanitary fittings							
Acrylic reinforced bath 1700mm long complete with chromium-plated grip handles, 40mm waste fitting, overflow, chain, plug bath panels, 20mm chromium-plated taps	1	nr	2.75	42.63	186.65	34.39	263.67
Vitreous china wash basin size 560 × 430mm, complete with 32mm waste fitting, overflow,chain, stay and plug, pair 13mm chromium-plated easy clean pillar taps, cast iron cantilever brackets and pedestal	1	nr	2.20	34.10	131.51	24.84	190.45
Carried forward			19.81	307.06	424.41	109.72	841.19

	Qty	Unit	Labour hours	Hours cost £	Materials £	O & P £	Total £
Brought forward			19.81	307.06	424.41	109.72	841.19
Shower cubicle size 788 × 842 × 2115mm high, anodised aluminium frame and safety glass, complete with 8.15kW instant electric shower with adjustable spray handset, slide bar and hose, soap dish and chromium-plated waste fittings	1	nr	3.80	58.90	751.33	121.53	931.76
Total			23.61	365.96	1,175.74	231.25	1,772.95

SUMMARY OF BATHROOM PROJECT COSTS

	Labour hours	Hours cost £	Materials £	O & P £	Total £
Bath with hand shower, WC and lavatory basin	19.94	309.07	691.80	150.13	1,151.00
Bath with hand shower, WC, lavatory basin and bidet	24.20	361.40	975.97	200.61	1,537.98
Bath with hand shower, lavatory basin and bidet	21.93	339.92	752.53	163.87	1,256.32
Bath with hand shower and lavatory basin	18.23	282.57	480.32	114.43	877.32
Bath with hand shower, WC, lavatory basin and shower cubicle	26.70	413.85	1,470.12	282.60	2,166.57
Bath with hand shower, WC, lavatory basin, shower cubicle and bidet	30.98	480.20	1,772.08	337.84	2,590.12
Bath with hand shower, lavatory basin, shower cubicle and bidet	27.49	426.10	1,632.11	308.73	2,366.94
Bath with hand shower, lavatory basin and shower cubicle	23.61	365.96	1,250.61	242.49	1,859.06
Bath, WC and lavatory basin	20.70	320.86	627.79	142.30	1,090.95
Bath, WC and lavatory basin and bidet	24.20	375.11	915.55	193.60	1,484.26
Bath, lavatory basin and bidet	22.22	344.42	510.14	128.18	982.74
Bath and lavatory basin	18.23	282.57	419.08	105.25	806.90

	Labour hours	Hours cost £	Materials £	O & P £	Total £
Bath, WC, lavatory basin and shower cubicle	26.30	402.07	1,386.58	268.30	2,056.95
Bath, WC, lavatory basin shower cubicle and bidet	31.34	485.78	1,699.23	327.75	2,512.76
Bath, lavatory basin, shower cubicle and bidet	27.49	426.10	1,465.98	283.81	2,175.89
Bath, lavatory basin and shower cubicle	23.61	365.96	1,175.74	231.26	1,772.96

	Qty	Unit	Labour hours	Hours cost £	Materials £	O & P £	Total £
EXTERNAL WASTE SYSTEMS							
MuPVC WASTE PIPES							
External waste system with solvent-welded joints for one-storey house							
Take down existing external waste pipes, remove and prepare surfaces to receive new		Item	2.00	31.00	2.00	4.95	37.95
32mm diameter waste pipe, solvent-welded joints, clips at 500mm maximum centres, plugged to brickwork	1.00	m	0.25	3.88	3.29	1.07	8.24
extra over for bend	1	nr	0.24	3.72	2.00	0.86	6.58
Connection to back inlet, caulking bush	1	nr	0.15	2.33	3.78	0.92	7.02
40mm diameter waste pipe, solvent-welded joints, clips at 500mm maximum centres, plugged to brickwork	2.00	m	0.56	8.68	7.08	2.36	18.12
extra over for bend	2	nr	0.52	8.06	4.55	1.89	14.50
Connection to back inlet, caulking bush	2	nr	0.30	4.65	7.56	1.83	14.04
Total			4.02	62.31	30.26	13.89	106.46

	Qty	Unit	Labour hours	Hours cost £	Materials £	O & P £	Total £
External waste system with solvent-welded joints for two-storey house							
Take down existing external waste pipes, remove and prepare surfaces to receive new		Item	2.50	38.75	2.00	6.11	46.86
32mm diameter waste pipe, solvent-welded joints, clips at 500mm maximum centres, plugged to brickwork	1.00	m	0.25	3.88	3.02	1.03	7.93
extra over for bend	1	nr	0.24	3.72	1.84	0.83	6.39
Connection to back inlet, caulking bush	1	nr	0.15	2.33	3.50	0.87	6.70
40mm diameter waste pipe, solvent-welded joints, clips at 500mm maximum centres, plugged to brickwork	2.00	m	0.50	7.75	6.50	2.14	16.39
extra over for bend	2	nr	0.52	8.06	4.22	1.84	14.12
Connection to back inlet, caulking bush	2	nr	0.30	4.65	7.00	1.75	13.40
Total			4.46	69.13	28.08	14.58	111.79

	Qty	Unit	Labour hours	Hours cost £	Materials £	O & P £	Total £
External waste system with solvent-welded joints for three-storey house							
Take down existing external waste pipes, remove and prepare surfaces to receive new		Item	3.00	46.50	1.00	7.13	54.63
32mm diameter waste pipe, solvent-welded joints, clips at 500mm maximum centres, plugged to brickwork	2.00	m	0.50	7.75	6.04	2.07	15.86
extra over for bend	2	nr	0.48	7.44	3.68	1.67	12.79
Connection to back inlet, caulking bush	2	nr	0.30	4.65	7.00	1.75	13.40
40mm diameter waste pipe, solvent-welded joints, clips at 500mm maximum centres, plugged to brickwork	3.00	m	0.84	13.02	9.75	3.42	26.19
extra over for bend	3	nr	0.78	12.09	6.33	2.76	21.18
Connection to back inlet, caulking bush	3	nr	0.45	6.98	10.50	2.62	20.10
Total			6.35	98.43	44.30	21.41	164.13

	Qty	Unit	Labour hours	Hours cost £	Materials £	O & P £	Total £
External waste system with push-fit joints for one-storey house							
Take down existing external waste pipes, remove and prepare surfaces to receive new		Item	2.00	31.00	1.00	4.80	36.80
32mm diameter waste pipe, push-fit joints, clips at 500mm maximum centres, plugged to brickwork	1.00	m	0.20	3.10	1.09	0.63	4.82
extra over for bend	1	nr	0.18	2.79	0.92	0.56	4.27
Connection to back inlet, caulking bush	1	nr	0.15	2.33	3.50	0.87	6.70
40mm diameter waste pipe, push-fit joints, clips at 500mm maximum centres, plugged to brickwork	2.00	m	0.48	7.44	3.68	1.67	12.79
extra over for bend	2	nr	0.40	6.20	1.96	1.22	9.38
Connection to back inlet, caulking bush	2	nr	0.30	4.65	7.00	1.75	13.40
Total			3.71	57.51	19.15	11.50	88.15

	Qty	Unit	Labour hours	Hours cost £	Materials £	O & P £	Total £
External waste system with push-fit joints for two-storey house							
Take down existing external waste pipes, remove and prepare surfaces to receive new		Item	2.50	38.75	1.00	5.96	45.71
32mm diameter waste pipe, push-fit joints, clips at 500mm maximum centres, plugged to brickwork	1.00	m	0.20	3.10	1.09	0.63	4.82
extra over for bend	1	nr	0.18	2.79	0.92	0.56	4.27
Connection to back inlet, caulking bush	1	nr	0.15	2.33	3.50	0.87	6.70
40mm diameter waste pipe, push-fit joints, clips at 500mm maximum centres, plugged to brickwork	2.00	m	0.48	7.44	3.68	1.67	12.79
extra over for bend	2	nr	0.40	6.20	1.96	1.22	9.38
Connection to back inlet, caulking bush	2	nr	0.30	4.65	7.00	1.75	13.40
Total			4.21	65.26	19.15	12.66	97.07

	Qty	Unit	Labour hours	Hours cost £	Materials £	O & P £	Total £
External waste system with push-fit joints for three-storey house							
Take down existing external waste pipes, remove and prepare surfaces to receive new		Item	3.00	46.50	1.00	7.13	54.63
32mm diameter waste pipe, push-fit joints, clips at 500mm maximum centres, plugged to brickwork	2.00	m	0.40	6.20	2.18	1.26	9.64
extra over for bend	2	nr	0.36	5.58	1.84	1.11	8.53
Connection to back inlet, caulking bush	2	nr	0.30	4.65	7.00	1.75	13.40
40mm diameter waste pipe, push-fit joints, clips at 500mm maximum centres, plugged to brickwork	3.00	m	0.72	11.16	5.52	2.50	19.18
extra over for bend	3	nr	0.60	9.30	2.94	1.84	14.08
Connection to back inlet, caulking bush	3	nr	0.45	6.98	10.50	2.62	20.10
Total			5.83	90.37	30.98	18.20	139.55

	Qty	Unit	Labour hours	Hours cost £	Materials £	O & P £	Total £
PVC-U SOIL PIPES							
External soil system with solvent-welded joints for one-storey house							
Take down existing soil pipes, remove and prepare surfaces to receive new		Item	4.00	62.00	1.00	9.45	72.45
110mm diameter soil pipe, solvent-welded joints, holder-bats at 500mm maximum centres, plugged	3.30	m	1.25	19.38	38.05	8.61	66.04
extra over for offset	1	nr	0.40	6.20	11.18	2.61	19.99
Connection to vitrified clayware drain	1	nr	0.40	6.20	17.32	3.53	27.05
Total			6.05	93.78	67.55	24.20	185.52

	Qty	Unit	Labour hours	Hours cost £	Materials £	O & P £	Total £
External soil system with solvent-welded joints for two-storey house							
Take down existing soil pipes, remove and prepare surfaces to receive new		Item	4.00	62.00	1.00	9.45	72.45
110mm diameter soil pipe, solvent-welded joints, holder-bats at 500mm maximum centres, plugged	6.60	m	2.50	38.75	76.10	17.23	132.08
Carried forward			6.50	100.75	77.10	26.68	204.53

	Qty	Unit	Labour hours	Hours cost £	Materials £	O & P £	Total £
Brought forward			6.50	100.75	77.10	26.68	204.53
extra over for							
offset	1	nr	0.40	6.20	11.18	2.61	19.99
branch	1	nr	0.40	6.20	19.31	3.83	29.34
Connection to vitrified clayware drain	1	nr	0.40	6.20	17.32	3.53	27.05
Total			7.70	119.35	124.91	36.64	280.90

	Qty	Unit	Labour hours	Hours cost £	Materials £	O & P £	Total £
External soil system with solvent-welded joints for three-storey house							
Take down existing soil pipes, remove and prepare surfaces to receive new		Item	6.00	93.00	1.00	14.10	108.10
110mm diameter soil pipe, solvent-welded joints, holder-bats at 500mm maximum centres, plugged	9.30	m	3.53	54.72	107.23	24.29	186.24
extra over for							
offset	1	nr	0.40	6.20	11.18	2.61	19.99
branch	2	nr	0.80	12.40	38.62	7.65	58.67
Connection to vitrified clayware drain	1	nr	0.40	6.20	17.32	3.53	27.05
Total			11.13	172.52	175.35	52.18	400.04

	Qty	Unit	Labour hours	Hours cost £	Materials £	O & P £	Total £
CAST IRON SOIL PIPES							
External soil system with flexible joints for one-storey house							
Take down existing soil pipes, remove and prepare surfaces to receive new		Item	4.00	62.00	1.00	9.45	72.45
100mm diameter soil pipe, flexible joints, holder-bats at 500mm maximum centres, plugged	3.30	m	2.15	33.33	97.25	19.59	150.16
extra over for offset	1	nr	0.55	8.53	17.08	3.84	29.45
Connection to vitrified clayware drain	1	nr	0.60	9.30	12.39	3.25	24.94
Total			7.30	113.15	127.72	36.13	277.00

	Qty	Unit	Labour hours	Hours cost £	Materials £	O & P £	Total £
External soil system with flexible joints for two-storey house							
Take down existing soil pipes, remove and prepare surfaces to receive new		Item	4.00	62.00	1.00	9.45	72.45
100mm diameter soil pipe, flexible joints, holder-bats at 500mm maximum centres, plugged	6.60	m	4.29	66.50	194.50	39.15	300.14
Carried forward			8.29	128.50	195.50	48.60	372.59

	Qty	Unit	Labour hours	Hours cost £	Materials £	O & P £	Total £
Brought forward			8.29	128.50	195.50	48.60	372.59
extra over for							
offset	1	nr	0.55	8.53	17.08	3.84	29.45
branch	1	nr	0.60	9.30	21.05	3.96	30.34
Connection to vitrified clayware drain	1	nr	0.60	9.30	12.39	3.25	24.94
Total			10.04	155.63	246.02	59.65	457.32

	Qty	Unit	Labour hours	Hours cost £	Materials £	O & P £	Total £
External soil system with flexible joints for three-storey house							
Take down existing soil pipes, remove and prepare surfaces to receive new		Item	6.00	93.00	1.00	14.10	108.10
100mm diameter soil pipe, flexible joints, holder-bats at 500mm maximum centres, plugged	9.90	m	6.05	93.78	274.07	55.18	423.02
extra over for							
offset	1	nr	0.55	8.53	17.08	3.84	29.45
branch	2	nr	1.20	18.60	42.10	9.11	69.81
Connection to vitrified clayware drain	1	nr	0.60	9.30	12.39	3.25	24.94
Total			14.40	223.20	346.64	85.48	655.32

SUMMARY OF EXTERNAL WASTE PROJECT COSTS

	Labour hours	Hours cost £	Materials £	O & P £	Total £
MuPVC waste system with solvent-welded joints					
One-storey house	4.02	62.31	30.26	13.89	106.46
Two-storey house	4.46	64.67	27.08	13.76	105.51
Three-storey house	6.35	98.43	44.30	21.41	164.14
MuPVC waste system with push-fit joints					
One-storey house	3.71	57.51	19.15	11.50	88.16
Two-storey house	4.21	65.26	19.15	12.66	97.07
Three-storey house	5.83	90.37	30.98	18.20	139.55
PVC-U soil system with solvent-welded joints					
One-storey house	6.05	93.78	67.55	24.20	185.53
Two-storey house	7.70	119.35	124.91	36.64	280.90
Three-storey house	11.13	172.52	175.35	52.18	400.05
Cast iron soil system with flexible joints					
One-storey house	7.30	113.15	246.02	36.13	277.00
Two-storey house	10.04	155.63	246.02	59.65	457.32
Three-storey house	14.40	223.20	346.64	85.48	655.32

	Qty	Unit	Labour hours	Hours cost £	Materials £	O & P £	Total £
CENTRAL HEATING SYSTEMS							
Central heating system for one-storey house, overall size 7 × 6m							
15mm diameter copper pipe, capillary joints and fittings, clips at 1250mm maximum centres	66	m	13.20	204.60	120.12	48.71	373.43
extra over for							
elbow	18	nr	3.24	50.22	18.00	10.23	78.45
tee	13	nr	2.86	44.33	13.00	8.60	65.93
Connection to heating boiler, 15mm diameter	4	nr	1.12	17.36	12.96	4.55	34.87
Connection to storage tank, 15mm diameter	2	nr	0.56	8.68	10.40	2.86	21.94
Break into existing copper pipe, insert 15m	1	nr	0.90	13.95	3.03	2.55	19.53
Gas-fired wall-mounted central combination boiler, 30,400 to 81,900 Btu, f	1	nr	4.80	74.40	694.29	115.30	883.99
Pressed steel radiator fixed to brickwork with concealed brackets, size							
single, 600 ×	1	nr	1.35	20.93	45.17	9.91	76.01
double, 600 ×	2	nr	2.50	38.75	115.42	23.13	177.30
double, 600 ×	2	nr	3.00	46.50	247.96	44.17	338.63
Chromium-plated radiator valve, 15mm ×	5	nr	1.50	23.25	98.65	18.29	140.19
Carried forward			35.03	542.97	1,379.00	288.29	2,210.26

	Qty	Unit	Labour hours	Hours cost £	Materials £	O & P £	Total £
Brought forward			35.03	542.97	1,379.00	228.29	2,210.26
Lockshield radiator valve, 15mm × 1/2in	5	nr	1.50	23.25	35.30	8.78	67.33
Galvanised steel tank ref. T25/1, 36 litres	1	nr	0.65	10.08	66.98	11.56	88.61
Hole through existing 100mm thick plastered block wall for two small pipes and make good	1	nr	0.40	6.20	0.00	0.93	7.13
Hole through existing stud partition plasterboard and skim both sides for two small pipes and make	3	nr	0.60	9.30	0.00	1.40	10.70
Hole through existing plasterboard and skim ceiling for two small pipes and make good	1	nr	0.20	3.10	0.00	0.47	3.57
Total			38.38	594.90	1,481.28	251.42	2,387.60

	Qty	Unit	Labour hours	Hours cost £	Materials £	O & P £	Total £
Central heating system for one-storey house, overall size 8 × 7m							
15mm diameter copper pipe, capillary joints and fittings, clips at 1250mm maximum centres	72	m	14.40	223.20	131.04	53.14	407.38
Carried forward			14.40	223.20	131.04	53.14	407.38

	Qty	Unit	Labour hours	Hours cost £	Materials £	O & P £	Total £
Brought forward			14.40	223.20	131.04	53.14	407.38
extra over for							
elbow	18	nr	3.24	50.22	18.00	10.23	78.45
tee	13	nr	2.86	44.33	13.00	8.60	65.93
Connection to heating boiler, 15mm diameter pipe	4	nr	1.12	17.36	12.96	4.55	34.87
Connection to storage tank, 15mm diameter pipe	2	nr	0.56	8.68	10.40	2.86	21.94
Break into existing copper pipe, insert 15mm tee	1	nr	0.90	13.95	3.03	2.55	19.53
Gas-fired wall-mounted central combination boiler, 30,400 to 81,900 Btu, fan flue	1	nr	4.80	74.40	694.29	115.30	883.99
Pressed steel radiator fixed to brickwork with concealed brackets, size							
single, 600 × 1400mm	1	nr	1.35	20.93	45.17	9.91	76.01
double, 600 × 1000mm	3	nr	3.75	58.13	173.31	34.72	266.15
double, 600 × 1800mm	2	nr	3.00	46.50	247.96	44.17	338.63
Chromium-plated radiator valve, 15mm × 1/2in	6	nr	1.80	27.90	118.38	21.94	168.22
Lockshield radiator valve, 15mm × 1/2in	6	nr	1.80	27.90	42.36	10.54	80.80
Galvanised steel tank ref. T25/1, 36 litres	1	nr	0.65	10.08	66.98	11.56	88.61
Carried forward			40.23	623.57	1,576.88	330.07	2,530.52

	Qty	Unit	Labour hours	Hours cost £	Materials £	O & P £	Total £
Brought forward			40.23	623.57	1,576.88	330.07	2,530.52
Hole through existing 100mm thick plastered block wall for two pipes, make good	1	nr	0.40	6.20	0.00	0.93	7.13
Hole through existing stud partition plasterboard and skim both sides for two small pipes and make good	3	nr	0.60	9.30	0.00	1.40	10.70
Hole through existing plasterboard and skim ceiling for two small pipes and make good	1	nr	0.20	3.10	0.00	0.47	3.57
Total			41.43	642.17	1,576.88	332.86	2,551.91

	Qty	Unit	Labour hours	Hours cost £	Materials £	O & P £	Total £
Central heating system for one-storey house, overall size 9 × 7m							
15mm diameter copper pipe, capillary joints and fittings, clips at 1250mm maximum centres	77	m	15.40	238.70	140.14	56.83	435.67
extra over for							
elbow	18	nr	3.24	50.22	18.00	10.23	78.45
tee	13	nr	2.86	44.33	13.00	8.60	65.93
Connection to heating boiler, 15mm diameter pipe	4	nr	1.12	17.36	12.96	4.55	34.87
Carried forward			22.62	350.61	184.10	80.21	614.92

	Qty	Unit	Labour hours	Hours cost £	Materials £	O & P £	Total £
Brought forward			22.62	350.61	184.10	80.21	614.92
Connection to storage tank, 15mm diameter pipe	2	nr	0.56	8.68	10.40	2.86	21.94
Break into existing copper pipe, insert 15mm tee	1	nr	0.90	13.95	3.03	2.55	19.53
Gas-fired wall-mounted central combination boiler, 35,500-95,000 Btu, condensing 1	1	nr	4.80	74.40	745.34	122.96	942.70
Pressed steel radiator fixed to brickwork with concealed brackets, size							
single, 600 × 1400mm	1	nr	1.35	20.93	45.17	9.91	76.01
double, 600 × 1000mm	4	nr	5.00	77.50	230.84	46.25	354.59
double, 600 × 1800mm	2	nr	3.00	46.50	247.96	44.17	338.63
Chromium-plated radiator valve, 15mm × 1/2in	7	nr	2.10	32.55	138.11	25.60	196.26
Lockshield radiator valve, 15mm × 1/2in	7	nr	2.10	32.55	49.42	12.30	94.27
Galvanised steel tank ref. T25/1, 36 litres	1	nr	0.65	10.08	66.98	11.56	88.61
Hole through existing 100mm thick plastered block wall for two small pipes and make good	1	nr	0.40	6.20	0.00	0.93	7.13
Carried forward			43.48	673.94	1,721.35	359.30	2,754.59

	Qty	Unit	Labour hours	Hours cost £	Materials £	O & P £	Total £
Brought forward			43.48	673.94	1,721.35	359.30	2,754.59
Hole through existing stud partition plasterboard and skim both sides for two small pipes and make	3	nr	0.60	9.30	0.00	1.40	10.70
Hole through existing plasterboard and skim ceiling for two small pipes and make good	1	nr	0.20	3.10	0.00	0.47	3.57
Total			44.28	686.34	1,721.35	361.16	2,768.85

	Qty	Unit	Labour hours	Hours cost £	Materials £	O & P £	Total £
Central heating system for two-storey house, overall size 7 × 7m							
15mm diameter copper pipe, capillary joints and fittings, clips at 1250mm maximum centres	113	m	22.60	350.30	205.56	83.38	639.24
extra over for							
elbow	29	nr	5.22	80.91	29.00	16.49	126.40
tee	15	nr	3.30	51.15	15.00	9.92	76.07
Connection to heating boiler, 15mm diameter	4	nr	1.12	17.36	12.96	4.55	34.87
Connection to storage tank, 15mm diameter	2	nr	0.56	8.68	10.40	2.86	21.94
Carried forward			32.80	508.40	272.92	117.20	898.52

	Qty	Unit	Labour hours	Hours cost £	Materials £	O & P £	Total £
Brought forward			32.80	508.40	272.92	117.20	898.52
Break into existing copper pipe, insert 15mm tee	1	nr	0.90	13.95	3.03	2.55	19.53
Gas-fired wall-mounted central combination boiler, 35,500-95,000 Btu, fan flue	1	nr	4.80	74.40	745.34	122.96	942.70
Pressed steel radiator fixed to brickwork with concealed brackets, size							
single, 600 × 1400mm	1	nr	1.35	20.93	45.17	9.91	76.01
double, 600 × 1000mm	5	nr	6.25	96.88	288.55	57.81	443.24
double, 600 × 1800mm	2	nr	3.00	46.50	247.96	44.17	338.63
Chromium-plated radiator valve, 15mm × 1/2in	8	nr	2.40	37.20	157.84	29.26	224.30
Lockshield radiator valve, 15mm × 1/2in	8	nr	2.40	37.20	56.48	14.05	107.73
Galvanised steel tank ref. T25/1, 86 litres	1	nr	0.75	11.63	82.09	14.06	107.77
Hole through existing 100mm thick plastered block wall for two small pipes and make good	2	nr	0.80	12.40	0.00	1.86	14.26
Hole through existing stud partition plasterboard and skim both sides for two small pipes and make good	5	nr	1.00	15.50	0.00	2.33	17.83
Carried forward			56.45	874.98	1,899.38	416.15	3,190.51

	Qty	Unit	Labour hours	Hours cost £	Materials £	O & P £	Total £
Brought forward			56.45	874.98	1,899.38	416.15	3,190.51
Hole through existing plasterboard and skim ceiling for two small pipes and make good	2	nr	0.40	6.20	0.00	0.93	7.13
Hole through existing softwood flooring for two small pipes, make good	1	nr	0.20	3.10	0.00	0.47	3.57
Total			57.05	884.28	1,899.38	417.55	3,201.21

	Qty	Unit	Labour hours	Hours cost £	Materials £	O & P £	Total £
Central heating system for two-storey house, overall size 8 × 7m							
15mm diameter copper pipe, capillary joints and fittings, clips at 1250mm maximum centres	116	m	23.20	359.60	211.12	85.61	656.33
extra over for							
elbow	29	nr	5.22	80.91	29.00	16.49	126.40
tee	15	nr	3.30	51.15	15.00	9.92	76.07
Connection to heating boiler, 15mm diameter pipe	4	nr	1.12	17.36	12.96	4.55	34.87
Connection to storage tank, 15mm diameter pipe	2	nr	0.56	8.68	10.40	2.86	21.94
Break into existing copper pipe, insert 15mm tee	1	nr	0.90	13.95	3.03	2.55	19.53
Carried forward			34.30	531.65	281.51	121.97	935.13

	Qty	Unit	Labour hours	Hours cost £	Materials £	O & P £	Total £
Brought forward			34.30	531.65	281.51	121.97	935.13
Gas-fired wall-mounted central combination boiler, 22,500-92,800 Btu, condensing flue	1	nr	4.80	74.40	872.77	142.08	1,089.25
Pressed steel radiator fixed to brickwork with concealed brackets, size							
single, 600 × 1400mm	1	nr	2.70	41.85	90.34	19.83	152.02
double, 600 × 1000mm	5	nr	6.25	96.88	288.55	57.81	443.24
double, 600 × 1800mm	2	nr	3.00	46.50	247.96	44.17	338.63
Chromium-plated radiator valve, 15mm × 1/2in	9	nr	2.70	41.85	175.57	32.61	250.03
Lockshield radiator valve, 15mm × 1/2in	9	nr	2.70	41.85	63.54	15.81	121.20
Galvanised steel tank ref. T25/1, 86 litres	1	nr	0.75	11.63	82.09	14.06	107.77
Hole through existing 100mm thick plastered block wall for two small pipes and make good	2	nr	0.80	12.40	0.00	1.86	14.26
Hole through existing stud partition plasterboard and skim both sides for two small pipes and make good	5	nr	1.00	15.50	0.00	2.33	17.83
Hole through existing plasterboard and skim ceiling for two small pipes and make good	2	nr	0.40	6.20	0.00	0.93	7.13
Carried forward			59.40	920.70	2,102.33	453.45	3,476.48

	Qty	Unit	Labour hours	Hours cost £	Materials £	O & P £	Total £
Brought forward			59.40	920.70	2,102.33	453.45	3,476.48
Hole through existing softwood flooring for two small pipes, make good	1	nr	0.20	3.10	0.00	0.47	3.57
Total			59.60	923.80	2,102.33	453.92	3,480.05

	Qty	Unit	Labour hours	Hours cost £	Materials £	O & P £	Total £
Central heating system for two-storey house, overall size 9 × 8m							
15mm diameter copper pipe, capillary joints and fittings, clips at 1250mm maximum centres	127	m	25.40	393.70	217.17	91.63	702.50
extra over for							
elbow	29	nr	5.22	80.91	14.50	14.31	109.72
tee	15	nr	3.30	51.15	10.50	9.25	70.90
Connection to heating boiler, 15mm diameter pipe	4	nr	1.12	17.36	12.00	4.40	33.76
Connection to storage tank, 15mm diameter pipe	2	nr	0.56	8.68	10.00	2.80	21.48
Break into existing copper pipe, insert 15mm tee	1	nr	0.90	13.95	2.60	2.48	19.03
Gas-fired wall-mounted central combination boiler, 22,500-92,800 Btu, condensing i	1	nr	4.80	74.40	815.68	133.51	1,023.59
Carried forward			41.30	640.15	1,082.45	258.39	1,980.99

	Qty	Unit	Labour hours	Hours cost £	Materials £	O & P £	Total £
Brought forward			41.30	640.15	1,082.45	258.39	1,980.99
Pressed steel radiator fixed to brickwork with concealed brackets, size							
single, 600 × 1400mm	2	nr	2.70	41.85	90.34	19.83	152.02
double, 600 × 1000mm	6	nr	5.00	77.50	346.26	63.56	487.32
double, 600 × 1800mm	2	nr	3.00	46.50	247.96	44.17	338.63
Chromium-plated radiator valve, 15mm × 1/2in	10	nr	3.00	46.50	197.30	36.57	280.37
Lockshield radiator valve, 15mm × 1/2in	10	nr	3.00	46.50	70.60	17.57	134.67
Galvanised steel tank ref. T25/1, 86 litres	1	nr	0.75	11.63	82.09	14.06	107.77
Hole through existing 100mm thick plastered block wall for two small pipes and make good	2	nr	0.80	12.40	0.00	1.86	14.26
Hole through existing stud partition plasterboard and skim both sides for two small pipes and make good	5	nr	1.00	14.50	0.00	2.18	16.68
Hole through existing plasterboard and skim ceiling for two small pipes and make good	2	nr	0.40	6.20	0.00	0.93	7.13
Carried forward			60.95	943.73	2,117.00	459.11	3,519.83

	Qty	Unit	Labour hours	Hours cost £	Materials £	O & P £	Total £
Brought forward			60.95	943.73	2,117.00	459.11	3,519.83
Hole through existing softwood flooring for two small pipes, make good	1	nr	0.20	3.10	0.10	0.48	3.68
Total			61.15	946.83	2,117.10	459.59	3,523.51

	Qty	Unit	Labour hours	Hours cost £	Materials £	O & P £	Total £
Central heating system for three-storey house, overall size 8 × 8m							
15mm diameter copper pipe, capillary joints and fittings, clips at 1250mm maximum centres	202	m	40.40	626.20	317.64	141.58	1,085.42
extra over for							
elbow	41	nr	7.38	114.39	41.00	23.31	178.70
tee	25	nr	5.50	85.25	25.00	16.54	126.79
Connection to heating boiler, 15mm diameter pipe	4	nr	1.12	17.36	121.96	20.90	160.22
Connection to storage tank, 15mm diameter pipe	2	nr	0.56	8.68	10.04	2.81	21.53
Break into existing copper pipe, insert 15mm tee	1	nr	0.90	13.95	3.03	2.55	19.53
Gas-fired wall-mounted central combination boiler, 25,900-93,800 Btu, condensing	1	nr	4.80	74.40	982.73	158.57	1,215.70
Carried forward			60.66	940.23	1,501.40	366.24	2,807.87

	Qty	Unit	Labour hours	Hours cost £	Materials £	O & P £	Total £
Brought forward			60.66	940.23	1,501.40	366.24	2,807.87
Pressed steel radiator fixed to brickwork with concealed brackets, size							
single, 600 × 1400mm	2	nr	2.70	41.85	90.34	19.83	152.02
double, 600 × 1000mm	7	nr	8.75	135.63	403.97	80.94	620.53
double, 600 × 1800mm	3	nr	4.50	69.75	431.94	75.25	576.94
Chromium-plated radiator valve, 15mm × 1/2in	12	nr	3.60	55.80	217.03	40.92	313.75
Lockshield radiator valve, 15mm × 1/2in	12	nr	3.60	55.80	84.72	21.08	161.60
Galvanised steel tank ref. T25/1, 86 litres	1	nr	0.75	11.63	66.98	11.79	90.40
Hole through existing 100mm thick plastered block wall for two small pipes and make good	2	nr	0.80	12.40	0.00	1.86	14.26
Hole through existing stud partition plasterboard and skim both sides for two small and make good	9	nr	1.80	27.90	0.00	4.19	32.09
Hole through existing plasterboard and skim ceiling for two small pipes and make good	3	nr	0.60	9.30	0.00	1.40	10.70
Carried forward			87.76	1,360.28	2,796.38	623.49	4,780.15

	Qty	Unit	Labour hours	Hours cost £	Materials £	O & P £	Total £
Brought forward			87.76	1,360.28	2,796.38	623.49	4,780.15
Hole through existing softwood flooring for two small pipes, make good	2	nr	0.40	6.20	0.20	0.96	7.36
Total			88.16	1,366.48	2,796.58	624.45	4,787.51

	Qty	Unit	Hours	Hours £	Mat'ls £	O & P £	Total £
Central heating system for three-storey house, overall size 9 × 8m							
15mm diameter copper pipe, capillary joints and fittings, clips at 1250mm maximum centres	217	m	41.40	641.70	394.94	155.50	1,192.14
extra over for							
elbow	41	nr	7.38	114.39	41.00	23.31	178.70
tee	25	nr	5.50	85.25	25.00	16.54	126.79
Connection to heating boiler, 15mm diameter pipe	4	nr	1.12	17.36	12.96	4.55	34.87
Connection to storage tank, 15mm diameter pipe	2	nr	0.56	8.68	10.04	2.81	21.53
Break into existing copper pipe, insert 15mm tee	1	nr	0.90	13.95	3.03	2.55	19.53
Gas-fired wall-mounted central combination boiler, 25,900-93,800 Btu, condensing t	1	nr	4.80	74.40	982.73	158.57	1,215.70
Carried forward			61.66	955.73	1,469.70	363.81	2,789.24

	Qty	Unit	Labour hours	Hours cost £	Materials £	O & P £	Total £
Brought forward			61.66	955.73	1,469.70	363.81	2,789.24
Pressed steel radiator fixed to brickwork with concealed brackets, size							
single, 600 × 1400mm	3	nr	4.05	62.78	135.51	29.74	228.03
double, 600 × 1000mm	8	nr	10.00	155.00	461.68	92.50	709.18
double, 600 × 1800mm	3	nr	4.50	69.75	431.94	75.25	576.94
Chromium-plated radiator valve, 15mm × 1/2in	14	nr	4.20	65.10	276.22	51.20	392.52
Lockshield radiator valve, 15mm × 1/2in	14	nr	4.20	65.10	98.84	24.59	188.53
Galvanised steel tank ref. T25/1, 86 litres	1	nr	0.75	11.63	66.98	11.79	90.40
Hole through existing 100mm thick plastered block wall for two small pipes and make good	2	nr	0.80	12.40	0.00	1.86	14.26
Hole through existing stud partition plasterboard and skim both sides for two small and make good	9	nr	1.80	27.90	0.00	4.19	32.09
Hole through existing plasterboard and skim ceiling for two small pipes and make good	3	nr	0.60	9.30	0.00	1.40	10.70
Carried forward			92.56	1,434.68	2,940.87	656.33	5,031.88

	Qty	Unit	Labour hours	Hours cost £	Materials £	O & P £	Total £
Brought forward			92.56	1,434.68	2,940.87	656.33	5.031.88
Hole through existing softwood flooring for two small pipes, make good	2	nr	0.40	6.20	0.00	0.93	7.13
Total			92.96	1,440.88	2,940.87	657.26	5,039.01

	Qty	Unit	Labour hours	Hours cost £	Materials £	O & P £	Total £
Central heating system for three-storey house, overall size 10 × 9m							
15mm diameter copper pipe, capillary joints and fittings, clips at 1250mm maximum centres	224	m	44.80	694.40	407.68	165.31	1,267.39
extra over for							
elbow	41	nr	7.38	114.39	41.00	23.31	178.70
tee	25	nr	5.50	85.25	25.00	16.54	126.79
Connection to heating boiler, 15mm diameter pipe	4	nr	1.12	17.36	12.96	4.55	34.87
Connection to storage tank, 15mm diameter pipe	2	nr	0.56	8.68	10.40	2.86	21.94
Break into existing copper pipe, insert 15mm tee	1	nr	0.90	13.95	3.03	2.55	19.53
Gas-fired wall-mounted central combination boiler, 25,900-93,800 Btu, condensing f	1	nr	4.80	74.40	982.73	158.57	1,215.70
Carried forward			65.06	1,008.43	1,482.80	373.68	2,864.91

	Qty	Unit	Labour hours	Hours cost £	Materials £	O & P £	Total £
Brought forward			65.06	1,008.43	1,482.80	373.68	2,864.91
Pressed steel radiator fixed to brickwork with concealed brackets, size							
single, 600 × 1400mm	3	nr	4.05	62.78	135.51	29.74	228.03
double, 600 × 1000mm	9	nr	11.25	174.38	519.39	104.06	797.83
double, 600 × 1800mm	3	nr	4.50	69.75	431.94	75.25	576.94
Chromium-plated radiator valve, 15mm × 1/2in	15	nr	4.50	69.75	295.95	54.86	420.56
Lockshield radiator valve, 15mm × 1/2in	15	nr	4.50	69.75	105.90	26.35	202.00
Galvanised steel tank ref. T25/1, 86 litres	1	nr	0.75	11.63	66.98	11.79	90.40
Hole through existing 100mm thick plastered block wall for two small pipes and make good	2	nr	0.80	12.40	0.00	1.86	14.26
Hole through existing stud partition plasterboard and skim both sides for two small pipes and make good	9	nr	1.80	27.90	0.00	4.19	32.09
Hole through existing plasterboard and skim ceiling for two small pipes and make good	3	nr	0.60	9.30	0.00	1.40	10.70
Carried forward			97.81	1,516.06	3,038.47	683.17	5,237.70

	Qty	Unit	Labour hours	Hours cost £	Materials £	O & P £	Total £
Brought forward			97.81	1,516.06	3,038.47	683.17	5,237.70
Hole through existing softwood flooring for two small pipes, make good	2	nr	0.40	6.20	0.00	0.93	7.13
Total			98.21	1,522.26	3,038.47	684.10	5,244.83

SUMMARY OF CENTRAL HEATING PROJECTS

	Labour hours	Hours cost £	Materials £	O & P £	Total £
One-storey house overall size					
7 × 6m (5 radiators)	38.38	594.90	1,481.28	251.42	2,387.60
8 × 7m (6 radiators)	41.43	642.17	1,576.88	332.66	2,551.91
9 × 7m (7 radiators)	44.28	686.34	1,721.35	361.16	2,768.85
Two-storey house overall size					
7 × 7m (8 radiators)	57.05	884.28	1,899.38	417.55	3,201.21
8 × 7m (9 radiators)	59.60	923.80	2,102.33	453.92	3,480.05
9 × 8m (10 radiators)	61.15	946.83	2,117.10	459.59	3,523.52
Three-storey house overall size					
8 × 8m (12 radiators)	88.16	1,366.48	2,796.58	624.46	4,787.52
9 × 8m (14 radiators)	92.96	1,440.88	2,940.87	657.26	5,039.01
10 × 9m (15 radiators)	98.21	1,522.26	3,038.47	684.11	5,244.84

	Qty	Unit	Labour hours	Hours cost £	Materials £	O & P £	Total £
HOT AND COLD WATER SUPPLY SYSTEMS							
One-storey house							
15mm diameter copper pipe, capillary joints and fittings, clips at 1250mm maximum centres	27.90	m	6.14	95.17	58.59	23.06	176.82
extra over for							
elbow	15	nr	2.70	41.85	15.00	8.53	65.38
tee	3	nr	0.66	10.23	3.00	1.98	15.21
Connection to tap, 15mm diameter pipe	3	nr	0.54	8.37	11.58	2.99	22.94
Connection to storage tank, 15mm diameter pipe	4	nr	1.12	17.36	21.60	5.84	44.80
22mm diameter copper pipe, capillary joints and fittings, clips at 1250mm maximum centres	15.20	m	3.65	56.58	47.27	15.58	119.42
extra over for							
elbow	6	nr	1.20	18.60	8.40	4.05	31.05
tee	3	nr	0.72	11.16	4.20	2.30	17.66
Connection to tap, 22mm diameter pipe	4	nr	0.80	12.40	39.64	7.81	59.85
Connection to storage tank, 22mm diameter pipe	2	nr	0.40	6.20	18.84	3.76	28.80
Plastic cistern with lid, reference PC 50, 227 litre:	1	nr	1.00	15.50	40.13	8.34	63.97
Carried forward			18.93	293.42	268.25	84.25	645.91

	Qty	Unit	Labour hours	Hours cost £	Materials £	O & P £	Total £
Brought forward			18.93	293.42	268.25	84.25	645.91
Gas fired wall-mounted combination boiler 30,400-81,900 Btu, fan flue	1	nr	4.80	74.40	694.29	115.30	883.99
Hole through existing stud partition plasterboard and skim both sides for two small pipes and make good	4	nr	2.40	37.20	0.00	5.58	42.78
Hole through existing plasterboard and skim ceiling for two small pipes and make good	2	nr	0.80	12.40	0.00	1.86	14.26
Hole through existing softwood flooring for two small pipes and make goo	2	nr	0.80	12.40	0.00	1.86	14.26
Total			27.73	429.82	962.54	208.85	1,601.20

	Qty	Unit	Hours	Hours £	Materials £	O & P £	Total £
Two-storey house							
15mm diameter copper pipe, capillary joints and fittings, clips at 1250mm maximum centres	30.60	m	6.23	96.57	64.26	24.12	184.95
extra over for							
elbow	15	nr	3.24	50.22	18.00	10.23	78.45
tee	3	nr	1.10	17.05	5.00	3.31	25.36
Connection to tap, 15mm diameter pipe	3	nr	0.54	8.37	11.58	2.99	22.94
Carried forward			11.11	172.21	98.84	40.66	311.70

	Qty	Unit	Labour hours	Hours cost £	Materials £	O & P £	Total £
Brought forward			11.11	172.21	98.84	40.66	311.70
Connection to storage tank, 15mm diameter pipe	4	nr	1.12	17.36	21.60	5.84	44.80
22mm diameter copper pipe, capillary joints and fittings, clips at 1250mm maximum centres	17.90	m	4.30	66.65	55.67	18.35	140.67
extra over for							
elbow	6	nr	1.60	24.80	11.20	5.40	41.40
tee	3	nr	0.72	11.16	7.00	2.72	20.88
Connection to tap, 22mm diameter pipe	4	nr	0.80	12.40	39.64	7.81	59.85
Connection to storage tank, 22mm diameter pipe	2	nr	0.40	6.20	18.84	3.76	28.80
Plastic cistern with lid, reference PC 50, 227 litre:	1	nr	1.00	15.50	40.13	8.34	63.97
Gas fired wall-mounted combination boiler 30,400-81,900 Btu, fan flue	1	nr	4.80	74.40	872.77	142.08	1,089.25
Hole through existing stud partition plasterboard and skim both sides for two small pipes and make good	4	nr	3.60	55.80	0.00	8.37	64.17
Hole through existing plasterboard and skim ceiling for two small pipes and make good	2	nr	1.20	18.60	0.00	2.79	21.39
Carried forward			30.65	475.08	1,165.69	246.12	1,886.88

	Qty	Unit	Labour hours	Hours cost £	Materials £	O & P £	Total £
Brought forward			30.65	475.08	1,165.69	246.12	1,886.88
Hole through existing softwood flooring for two small pipes and make goo	2	nr	1.20	18.60	0.00	2.79	21.39
Total			31.85	493.68	1,165.69	248.91	1,908.27

	Qty	Unit	Labour hours	Hours cost £	Materials £	O & P £	Total £
Three-storey house							
15mm diameter copper pipe, capillary joints and fittings, clips at 1250mm maximum centres	33.30	m	7.32	113.46	69.93	27.51	210.90
extra over for							
elbow	15	nr	3.96	61.38	22.00	12.51	95.89
tee	3	nr	1.76	27.28	8.00	5.29	40.57
Connection to tap, 15mm diameter pipe	3	nr	0.54	8.37	11.58	2.99	22.94
Connection to storage tank, 15mm diameter pipe	4	nr	1.12	17.36	21.60	5.84	44.80
22mm diameter copper pipe, capillary joints and fittings, clips at 1250mm maximum centres	20.60	m	4.94	76.57	64.07	21.10	161.74
extra over for							
elbow	6	nr	2.00	31.00	14.00	6.75	51.75
tee	3	nr	1.68	26.04	9.80	5.38	41.22
Carried forward			23.32	361.46	220.98	87.37	669.81

	Qty	Unit	Labour hours	Hours cost £	Materials £	O & P £	Total £
Brought forward			23.32	361.46	220.98	87.37	669.81
Connection to tap, 22mm diameter pipe	4	nr	0.80	12.40	39.64	7.81	59.85
Connection to storage tank, 22mm diameter pipe	2	nr	0.40	6.20	18.84	3.76	28.80
Plastic cistern with lid, reference PC 50, 227 litre:	1	nr	1.00	15.50	40.17	8.35	64.02
Gas fired wall-mounted combination boiler 25,900-93,800 Btu, condensing flı	1	nr	4.80	74.40	982.73	158.57	1,215.70
Hole through existing stud partition plasterboard and skim both sides for two small pipes and make good	4	nr	4.80	74.40	0.00	11.16	85.56
Hole through existing plasterboard and skim ceiling for two small pipes and make good	2	nr	1.60	24.80	0.00	3.72	28.52
Hole through existing softwood flooring for two small pipes and make goo	2	nr	1.60	24.80	0.00	3.72	28.52
Total			38.32	593.96	1,302.36	284.45	2,180.77

SUMMARY OF HOT AND COLD WATER PROJECTS

	Hours	Hours £	Materials £	O & P £	Total £
One-storey house	27.73	429.82	962.54	208.85	1,601.20
Two-storey house	31.85	493.68	1,165.69	248.91	1,908.28
Three-storey house	38.32	593.96	1,302.36	284.45	2,180.77

Part Three

BUSINESS MATTERS

Starting a business

Running a business

Taxation

Starting a business

Most small businesses come into being for one of two reasons – ambition or desperation! A person with genuine ambition for commercial success will never be completely satisfied until he has become self-employed and started his own business. But many successful businesses have been started because the proprietor was forced into this course of action because of redundancy.

Before giving up his job, the would-be businessman should consider carefully whether he has the required skills and the temperament to survive in the highly competitive self-employed market. Before commencing in business it is essential to assess the commercial viability of the intended business because it is pointless to finance a business that is not going to be commercially viable.

In the early stages it is important to make decisions such as: What exactly is the product being sold? What is the market view of that product? What steps are required before the developed product is first sold and where are those sales coming from?

As much information as possible should be obtained on how to run a business before taking the plunge. Sales targets should be set and it should be clearly established how those important first sales are obtained. Above all, do not underestimate the amount of time required to establish and finance a new business venture.

Whatever the size of the business it is important that you put in writing exactly what you are trying to do. This means preparing a business plan that will not only assist in establishing your business aims but is essential if you need to raise finance. The contents of a typical business plan are set out later. It is important to realise that you are not on your own and there are many contacts and advising agencies that can be of assistance.

Potential customers and trade contacts

Many persons intending to start a business in the construction industry will have already had experience as employees. Use all contacts to check the market, establish the sort of work that is available and the current charge-out rates.

In the domestic market, check on the competition for prices and services provided. Study advertisements for your kind of work and try to get firm promises of work before the start-up date.

Testing the market

Talk to as many traders as possible operating in the same field. Identify if the market is in the industrial, commercial, local government or in the domestic field. Talk to prospective customers and clients and consider how you can improve on what is being offered in terms of price, quality, speed, convenience, reliability and back-up service.

Business links

There is no shortage of information about the many aspects of starting and running your own business. Finance, marketing, legal requirements, developing your business idea and taxation matters are all the subject of a mountain of books, pamphlets, guides and courses so it should not be necessary to pay out a lot of money for this information. Indeed, the likelihood is that the aspiring businessman will be overwhelmed with information and will need professional guidance to reduce the risk of wasting time on studying unnecessary subjects.

Business Links are now well established and provide a good place to start for both information and advice. These organisations provide a 'one-stop-shop' for advice and assistance to owner-managed businesses. They will often replace the need to contact Training and Enterprise Councils (TECs) and many of the other official organisations listed below.
Point of contact: telephone directory for address.

Training and Enterprise Councils (TECs)

TECs are comprised of a board of directors drawn from the top men in local industry, commerce, education, trade unions etc., who, together with their staff and experienced business counsellors, assist both new and established concerns in all aspects of running a business. This takes the form of across-the-table advice and also hands-on assistance in management, marketing and finance if required. There are also training courses and seminars available in most areas together with the possibility of grants in some areas.
Point of contact: local Jobcentre or Citizens' Advice Bureau.

Banks

Approach banks for information about the business accounts and financial services that are available. Your local Business Link can advise on how best to find a suitable bank manager and inform you as to what the bank will require.

Shop around several banks and branches if you are not satisfied at first because managers vary widely in their views on what is a viable business proposition. Remember, most banks have useful free information packs to help business start-up.

Point of contact: local bank manager.

HM Inspector of Taxes

Make a preliminary visit to the local tax office enquiry counter for their publications on income tax and national insurance contributions.

SA/Bk 3	Self assessment. A guide to keeping records for the self employed
IR 15(CIS)	Construction Industry Tax Deduction Scheme
CWF1	Starting your own business,
IR 40(CIS)	Conditions for Getting a Sub-Contractor's Tax Certificate
NE1	PAYE for Employers (if you employ someone)
NE3	PAYE for new and small Employers
IR 56/N139	Employed or Self-Employed. A guide for tax and National Insurance
CA02	National Insurance contributions for self employed people with small earnings.

Remember, the onus is on the taxpayer, within three months, to notify the Inland Revenue that he is in business and failure to do so may result in the imposition of £100 penalty. Either send a letter or use the form provided at the back of the *'Starting your own business booklet'* to the Inland Revenue National Insurance Contributions Office and they will inform your local tax office of the change in your employment status.

Point of contact: telephone directory for address.

Inland Revenue National Insurance Contributions Office

Self Employment Services
Customer Accounts Section
Longbenton
Newcastle NE 98 1ZZ

Telephone the Call Centre on 0845 9154655 and ask for the following publications:

CWL2	Class 2 and Class 4 Contributions for the Self Employed
CA02	People with Small Earnings from Self-Employment
CA04	Direct Debit - The Easy Way to Pay. Class 2 and Class 3
CA07	Unpaid and Late Paid Contributions
and for Employers	
CWG1	Employer's Quick Guide to PAYE and NIC Contributions
CA30	Employer's Manual to Statutory Sick Pay

VAT

The VAT office also offer a number of useful publications, including;

700	The VAT Guide
700/1	Should I be Registered for VAT?
731	Cash Accounting
732	Annual Accounting
742	Land and Property

Information about the Cash Accounting Scheme and the introduction of annual VAT returns are dealt with later.
Point of contact: telephone directory for address.

Local authorities

Authorities vary in provisions made for small businesses but all have been asked to simplify and cut delays in planning applications. In Assisted Areas, rent-free periods and reductions in rates may be available on certain industrial and commercial properties. As a preliminary to either purchasing or renting business premises, the following booklets will be helpful:

Step by Step Guide to Planning Permission for Small Businesses, and *Business Leases and Security of Tenure*

Both are issued by the Department of Employment and are available at council offices, Citizens' Advice Bureaux and TEC offices. Some authorities run training schemes in conjunction with local industry and educational establishments.

Point of contact: usually the Planning Department - ask for the Industrial Development or Economic Development Officer.

Department of Trade and Industry

The services formally provided by the Department are now increasingly being provided by Business Link . The Department can still, however, provide useful information on available grants for start-ups.
Point of contact: telephone 0207-215 5000 and ask for the address and telephone number of the nearest DTI office and copies of their explanatory booklets.

Department of Transport and the Regions

Regulations are now in force relating to all forms of waste other than normal household rubbish. Any business that produces, stores, treats, processes, transports, recycles or disposes of such waste has a 'duty of care' to ensure it is properly discarded and dealt with.

Practical guidance on how to comply with the law (it is a criminal offence punishable by a fine not to) is contained in a booklet *Waste Management: The Duty of Care: A Code of Practice,* obtainable from HMSO Publication Centre, PO Box 276, London SW8 5DT. Telephone 0207-873 9090.

Accountant

The services of an accountant are to be strongly recommended from the beginning because the legal and taxation requirements start immediately and must be properly complied with if trouble is to be avoided later. A qualified accountant must be used if a limited company is being formed but an accountant will give advice on a whole range of business issues including book-keeping, tax planning and compliance to finance raising and will help in preparing annual accounts.

It is worth spending some time finding an accountant who has other clients in the same line of business and is able to give sound advice particularly on taxation and business finance and is not so overworked that damaging delays in producing accounts are likely to arise. Ask other traders whether they can recommend their own accountant. Visit more than one firm of accountants, ask about the fees they charge and how much the production of annual accounts and agreement with the Inland Revenue are likely to cost. A good accountant is worth every penny of his fees and will save you both money and worry.

Solicitor

Many businesses operate without the services of a solicitor but there are a number of occasions when legal advice should be sought. In particular, no-one should sign a lease of premises without taking legal advice because a business can encounter financial difficulty through unnoticed liabilities in its lease. Either an accountant or solicitor will help with drawing up a partnership agreement which all partnerships should have. A solicitor will also help to explain complex contractual terms and prepare draft contracts if the type of business being entered into requires them.

Insurance broker

Policies are available to cover many aspects of business including:

- employer's liability - compulsory if the business has employees
- public liability - essential in the construction industry
- motor vehicles
- theft of stock, plant and money
- fire and storm damage
- personal accident and loss of profits
- keyman cover.

Brokers are independent advisers who will obtain competitive quotations on your behalf. See more than one broker before making a decision - their advice is normally given free and without obligation.
Point of contact: telephone directory or write for a list of local members to:

The British Insurance Brokers' Association
Consumer Relations Department
BIBA House
14 Bevis Marks
London
EC3A 7NT (telephone: 0207-623 9043)
or contact
The Association of British Insurers
51 Gresham Street
London
EC2V 7HQ (telephone: 0207-600 3333)

The British Insurance Brokers' Association
Consumer Relations Department
BIBA House
14 Bevis Marks
London
EC3A 7NT (telephone: 0207-623 9043)
or contact
The Association of British Insurers
51 Gresham Street
London
EC2V 7HQ (telephone: 0207-600 3333)

who will supply free a package of very useful advice files specially designed for the small business.

The Health and Safety Executive

The Executive operates the legislation covering everyone engaged in work activities and has issued a very useful set of '*Construction Health Hazard Information Sheets*' covering such topics as handling cement, lead and solvents, safety in the use of ladders, scaffolding, hoists, cranes, flammable liquids, asbestos, roofs and compressed gases etc. A pack of these may be obtained free from your local HSE office or The Health & Safety Executive Central Office, Sheffield (telephone: (01142-892345) or HSE Publications (telephone: 01787-881165).

Business plan

As stated before, once the relevant information has been obtained it should be consolidated into a formal business plan. The complexity of the plan will depend in the main on the size and nature of the business concerned. Consideration should be given to the following points.

Objectives

It is important to establish what you are trying to achieve both for you and the business. A provider of finance may be particularly influenced by your ability to achieve short- and medium-term goals and may have confidence in continuing to provide finance for the business. From an individual point of view, it is important to establish goals because there is little point in having a business that only serves to achieve the expectations of others whilst not rewarding the would-be businessman.

History

If you already own an existing business then commentary on its existing background structure and history to date can be of assistance. There is no substitute for experience and any existing contacts you have in the construction industry will be of assistance to you. The following points should also be considered for inclusion:

- a brief history of the business identifying useful contacts made
- the development of the business, highlighting significant successes and their relevance to the future
- principal reasons for taking the decision to pursue this new venture
- details of present financing of the business.

Products or services

It is important to establish precisely what it is you are going to sell. Does the product or service have any unique qualities which gives it your advantages over competitors? For example, do you have an ability to react more quickly than your competitors and are you perceived to deliver a higher quality product or service?
A typical business plan would include:

- description of the main products and services
- statement of disadvantages and advising how they will be overcome
- details of new products and estimated dates of introduction
- profitability of each product
- details of research and development projects
- after-sales support.

Markets and marketing strategy

This section of the business plan should show that thought has been given to the potential of the product. In this regard it can often be useful to identify major competitors and make an overall assessment of their strengths and weaknesses, including the following:

- an overall assessment of the market, setting out its size and growth potential
- a statement showing your position within the market

- an identification of main customers and how they compare
- details of typical orders and buying habits
- pricing strategy
- anticipated effect on demand of pricing
- expectation of price movement
- details of promotions and advertising campaigns.

It is important to identify your customers and why they might buy from you. Those entering the domestic side of the business will need to think about the best way to reach potential customers. Are local word-of-mouth recommendations enough to provide reasonable work continuity. If not, what is the most effective method of advertising to reach your customer base?

Remember, advertising is costly. It is a waste of funds to place an advertisement in a paper circulating in areas A, B, C & D if the business only covers area A.

Research and development

If you are developing a product or a particular service, then an assessment should be made on what stage it is at and what further finance is required to complete it. It may also be useful to make an assessment on the vulnerability of the product or service to innovations being initiated by others.

Basis of operation

Detail what facilities you will require in order to carry on your trade in the form of property, working and storage areas, office space, etc. An assessment should also he made on the assistance you will require from others.

Your business plan might include:

- a layman's guide to the process or work
- details of facilities, buildings and plant
- key factors affecting production, such as yields and wastage
- raw material demand and usage.

Management

This section is one of the most important because it demonstrates the capability of the would-be businessman. The skills you need will cover production, marketing, finance and administration. In the early stages you may be able to do this yourself but as the business grows it may be required to develop a team to handle these matters. The following points should be considered for inclusion in the plan:

- set out age, experience and achievements
- state additional management requirements in the future and how they are to be met
- identify current weaknesses and how they will be overcome
- state remuneration packages and profit expectations
- give detailed CVs in appendices.

Advertising and retraining may be required in order to identify and provide suitable personnel where expertise and experience are lacking.

Financial information

It is important to detail, if any, the present financial position of your business and the budgeted profit and loss accounts, cash flows and balance sheets. These integrated forecasts should be prepared for the next twelve months at monthly intervals and annually for the following two years.

If the forecasts are to be reasonably accurate then the businessman must make some early decisions about:

- the premises where the business will be based, the initial repairs and alterations that might he required and an assessment of the total cost
- which plant, equipment and transport are needed, whether they are to be leased or purchased and what the cost will be?
- how much stock of materials, if any, should be carried? - the bare minimum only should be acquired, so reliable suppliers should be found
- what will be the weekly bills for overheads, wages and the proprietor's living costs?
- what type of work is going to be undertaken, and how much profit can realistically be obtained?
- how often are invoices to be presented?

Your business plan should include the following information:

- explanation of how sales forecasts are prepared
- levels of production
- details of major variable overheads and estimates
- assumptions in cash flow forecasting, inflation and taxation.

Finance required and its application

The financial details given above should produce an accurate assessment of the funds required to finance the business. It is important to distinguish between those items that require permanent finance and those that will eventually be converted to cash because it is not usually advisable to finance long-term assets with personal equity.

Working capital such as stock and debtors can usually be obtained by an overdraft arrangement but your accountant or bank will advise you on this.

Executive summary

Although it is prepared last, this summary will be the first part of your business plan. Remember that business plans are prepared for busy people and their decision on finance may be based solely on this section. It should cover two or three pages and deal with the most important aspects and opportunities in your plan. Here are some of the main headings:

- key strategies
- finance required and how it is to be used
- management experience
- anticipated returns and profits
- markets.

The appendices should include:

- CVs of key personnel
- organisation charts
- market studies
- product advertising literature
- professional references
- financial forecasts
- glossary of terms.
- anticipated returns and profits
- markets.

If you feel that any additional information should be provided in support of your proposal, then this is usually best included in the appendices.

Follow up

Please remember that once your plan is prepared, it is important to re-examine it regularly and update the forecasts and financial information. This is a working document and can be an important tool in running the business.

Sources of finance

Personal funds

Finance, like charity, often begins at home and a would-be businessman should make a realistic assessment of his net worth, including the value of his house after deducting the mortgage(s) outstanding on it, savings, any car or van owned and any sums which the family are prepared to contribute but deducting any private borrowings which will come due for payment. The whole of these funds may not be available (for instance, money which has been loaned to a friend or relative who is known to be unable to repay at the present time).

It may not be desirable that all capital should be put at risk on a business venture so the following should be established:

- how much cash you propose to invest in the business
- whether the family home will be made available for any business borrowing
- state total finance required
- how finance is anticipated being raised
- interest and security to be provided
- expected return on investment.

Whilst it may be wise not to pledge too much of the family assets, it has to be remembered that the bank will be looking closely at the degree to which the proprietor has committed himself to the venture and will not be impressed by an application for a loan where the applicant is prepared to risk only a small fraction of his own resources.

Having decided how much of his own funds to contribute, the businessman can now see the level of shortfall and consider how best to fill it. Consideration should be given to partners where the shortfall is large and particularly when there is a need for heavy investment in fixed assets, such as premises and capital equipment. It may be worthwhile starting a limited company with others also subscribing capital and to allow the banks to take security against the book debts.

Banks

The first outside source of money to which most businessmen turn is the bank and here are a few guidelines on approaching a bank manager:

- present your business plan to him; remember to use conservative estimates which tend to understate rather than overstate the forecast sales and profits
- know the figures in detail and do not leave it to your accountant to explain them for you. The bank manager is interested in the businessman not his advisers and will be impressed if the businessman demonstrates a grasp of the financing of his business
- understand the difference between short- and long-term borrowing
- ask about the Government Loan Guarantee Scheme if there is a shortage of security for loans. The bank may be able to assist, or depending on certain conditions being met, the Government may guarantee a certain percentage of the bank loan.

Remember the bank will want their money back, so bank borrowings are usually required to be secured by charges on business assets. In start-up situations, personal guarantees from the proprietors are normally required. Ensure that if these are given they are regularly reviewed to see if they are still required.

Enterprise Investment Scheme - business angels

If an outside investor is sought in a business he will probably wish to invest within the terms of the Enterprise Investment Scheme which enables him to gain income tax relief at 20% on the amount of his investment. Additionally, any investment can be used to defer capital gains tax. The rules are complex and professional advice should always be sought.

Hire purchase/leasing

It is not always necessary to purchase assets outright that are required for the business and leasing and hire purchase can often form an integral part of a business's medium-term finance strategy.

Venture capital

In addition, there are a number of other financial institutions in the venture capital market that can help well-established businesses, usually limited companies, who wish to expand. They may also assist well-conceived start-ups. They will provide a flexible package of equity and loan capital but only for large amounts, usually sums in excess of £150,000 and often £250,000.

Usually the deal involves the financial institution having a minority interest in the voting share capital and a seat on the board of the company. Arrangements for the eventual purchase of the shares held by the finance company by the private shareholders are also normally incorporated in the scheme.

The Royal Jubilee and Princes Trust

These trusts through the Youth Business Initiative provide bursaries of not more than £1,000 per individual to selected applicants who are unemployed and age 25 or over. Grants may be used for tools and equipment, transport, fees, insurance, instruction and training but not for working capital, rent and rates, new materials or stock. They operate through a local representative whose name and address may be ascertained by contacting the Prince's Youth Business.
Point of contact: telephone 0207-321 6500.

The Business Start-up Scheme

This is an allowance of £50 per week, in addition to any income made from your business, paid for twenty weeks. To qualify you must be at least 18 and under 65, work at least 36 hours per week in the business and have been unemployed for at least six months or fall into one of the other categories: disabled, ex-HMS or redundant.

The first step is to get the booklet on the subject from your local Jobcentre or TEC that includes details on how and where to apply. Once in receipt of the enterprise allowance, you will also have the benefit of advice and assistance from an experienced businessman from your TEC. All the initial counselling services and training courses are free.

Running a business

Many businesses are run without adequate information being available to check trend in their vital areas, e.g. marketing, money and managerial efficiency. It is essential to look critically at all aspects of the business in order to maximise profits and reduce inefficiency. Regular meaningful information is required on which management can concentrate. This will vary according to the proprietor's business but will often concentrate on debtors, creditors, cash, sales and orders.

Proprietors often have the feeling that the business should be 'doing better' but are unable to identify what is going wrong. Sometimes there is the worrying phenomenon of a steadily increasing work programme coupled with a persistently reducing bank balance or rising overdraft. Some useful ways of checking the position and of identifying problem areas are given below.

Marketing

Throughout his business life the entrepreneur should continuously study the methods and approach of his competitors. A shortcoming frequently found in ailing concerns is that the proprietor thinks he knows what his customers want better than they do.

The term 'market research' sounds both difficult and expensive but a very simple form of it can be done quite effectively by the businessman and his sales staff. Existing and prospective customers should be approached and asked what they want in terms of price, quality, design, payment terms, follow-up service, guarantees and services.

The initial approach might be by a leaflet or letter followed by a personal call. As an on-going part of management, all staff with customer contact should be encouraged to enquire about and record customer preferences, complaints, etc. and feed it back to management.

Other sources of information can be trade and business journals, trade exhibitions, suppliers and representatives from which information about trends, new techniques and products can be obtained and studied. Valuable information can also be gained from studying competitors and the following questions should be asked:

- what do they sell and at what prices?
- what inducements do they offer to their customers, e.g. credit facilities, guarantees, free offers and discounts?
- how do they reach their customers - local/national advertising, mail shots, salesmen, local radio and TV?
- what are the strongest aspects of their appeal to customers and
- have they any weaknesses?

The businessman should apply all the information gathered from customers and competitors to his own services with a view to making sure he is offering the right product at the right price in the most attractive way and in the most receptive market.

In a small business where the proprietor is also his own salesman he must give careful thought on how he can best present his product and himself. For instance, if he is working solely within the construction industry his main problems are likely to centre on getting a C1S6 Certificate and using trade contacts to get sub-contract work.

However, for those who serve the general public, presentation can be a vital element in getting work. The customer is looking for efficiency, reliability and honesty in a trader and quality, price and style in the product. To bring out these facets in discussion with a potential customer is a skilled task. A short course on marketing techniques could pay handsome dividends. The Business Link will give the names and addresses of such courses locally.

Financial control

Unfortunately, some unsuccessful firms do not seek financial advice until too late when the downward trend cannot be halted. Earlier attention to the problems may have saved some of them so it is important to recognise the tell-tale signs. There are some tests and checks that can be done quite easily.

Cash flow

Cash flow is the lifeblood of the business and more businesses fail through lack of cash than for any other reason. Cash is generated through the conversion of work into debtors and then into payment and also through the deferral of the payment of supplies for as long a period that can be negotiated. The objective must be to keep stock, work in progress, debts to a minimum and creditors to a maximum.

Debtor days

This is calculated by dividing your trade debtors by annual sales and multiplying by 365. This shows the number of days' credit being afforded to your customers and should be compared both with your normal trade terms and the previous month's figures. Normal procedures should involve the preparation of a monthly-aged list of debtors showing the name of the customer, the value and to which month it relates.

The oldest and largest debtors can be seen at a glance for immediate consideration of what further recovery action is needed. The list may also show over-reliance on one or two large customers or the need to stop supplying a particularly bad payer until his arrears have been reduced to an acceptable level. Consideration should be given to making up bills to a date before the end of the month and making sure the accounts are sent out immediately, followed by a statement four weeks later.

Consider giving discounts for prompt payment. If all else fails, and legal action for recovery is being contemplated, call at the County Court and ask for their leaflets.

Stock turn

The level of stock should be kept to a minimum and the number of days' stock can be calculated by dividing the stock by the annual purchases and multiplying by 365. A worsening trend on a month-by-month basis shows the need for action. It is important to regularly make a full inventory of all stock and dispose of old or surplus items for cash. A stock control procedure to avoid stock losses and to keep stock to a minimum should be implemented.

Profitability

Whilst cash is vital in the short-term, profitability is vital in the medium-term. The two key percentage figures are the gross profit percentage and the net profit percentage. Gross profit is calculated by deducting the cost of materials and direct labour from the sales figures whilst net profit is arrived at after deducting all overheads. Possible reasons for changes in the gross profit percentage are:

- not taking full account of increases in materials and wages in the pricing of jobs
- too generous discount terms being offered
- poor management, over-manning, waste and pilferage of materials
- too much down-time on equipment which is in need of replacement.

If net profit is deteriorating after the deduction of an appropriate reward for your own efforts, including an amount for your own personal tax liability, you should review each item of overhead expenditure in detail asking the following questions:

- can savings be made in non-productive staff?
- is sub-contracting possible and would it be cheaper?

- have all possible energy-saving methods been fully explored?
- do the company's vehicles spend too much time in the yard and can they be shared or their number reduced?
- is the expenditure on advertising producing sales - review in association with 'marketing' above?

Over-trading

Many inexperienced businessmen imagine that profitability equals money in the bank and in some cases, particularly where the receipts are wholly in cash, this may be the case. But often, increased business means higher stock inventories, extra wages and overheads, increased capital expenditure on premises and plant, all of which require short-term finance.

Additionally, if the debtors show a marked increase as the turnover rises, the proprietor may find to his surprise that each expansion of trade reduces rather than increases his cash resources and he is continually having to rely on extensions to his existing credit.

The business, which had enough funds for start-up, finds it does not have sufficient cash to run at the higher level of operation and the bank manager may he getting anxious about the increasing overdraft. It is essential for those who run a business that operates on credit terms to be aware that profitability does not necessarily mean increased cash availability. Regular monthly management information on marketing and finance as described in this chapter will enable over-trading to be recognised and remedial action to be taken early.

If the situation is appreciated only when the bank and other creditors are pressing for money, radical solutions may be necessary, such as bringing in new finance, sale and leaseback of premises, a fundamental change in the terms of trade or even selling out to a buyer with more resources. Professional help from the firm's accountant will be needed in these circumstances.

Break-even point

The costs of a business may be divided into two types - variable and fixed. *Variable costs* are those which increase or decrease as the volume of work goes up or down and include such items as materials used, direct labour and power machine tools. *Fixed costs* are not related to turnover and are sometimes called fixed overheads. They include rent, rates, insurance, heat and light, office salaries and plant depreciation. These costs are still incurred even though few or no sales are being made.

Many small businessmen run their enterprises from home using family labour as back-up; they mainly sell their own labour and buy materials and hire plant only as required. By these means they reduce their fixed costs to a minimum and start making profits almost immediately. However, larger firms that have business premises, perhaps a small workshop, an office and vehicles, need to know how much they have to sell to cover their costs and become profitable.

In the case of a new business it is necessary to estimate this figure but where annual accounts are available a break-even chart based on them can be readily prepared. Suppose the real or estimated figures (expressed in £000s) are:

	%	£
Sales	100	400
Variable costs	66	265
Gross profit	34	135
Fixed costs	13	50
Net profit	21	85

Break-even point = 50 divided by (1 less variable costs %) / sales

= 50 divided by (1 less 0.6625)
= 50 divided by 0.3375
= £148 (thousand)

In practice things are never quite as clear cut as the figures show, but nevertheless this is a very useful tool for assessing not only the break-even point but also the approximate amount of loss or profit arising at differing levels of turnover and also for considering pricing policy.

Taxation

The first decision usually required to be made from a taxation point of view is which trading entity to adopt. The options available are set out below.

Sole trader

A sole trader is a person who is in business on his own account. There is no statutory requirement to produce accounts nor is there a necessity to have them audited. A sole trader may, however, be required to register for PAYE and VAT purposes and maintain records so that Income Tax and VAT returns can be made. A sole trader is personally liable for all the liabilities of his business.

Partnership

A partnership is a collection of individuals in business on their own account and whose constitution is generally governed by the Partnership Act 1890. It is strongly recommended that a partnership agreement is also established to determine the commercial relationship between the individuals concerned.

The requirements in relation to accounting records and returns are similar to those of a sole trader and in general a partner's liability is unlimited.

Limited company

This is the most common business entity. Companies are incorporated under the Companies Act 1985 which requires that an annual audit is carried out for all companies with a turnover in excess of £1,000,000 or a review if the turnover is less than £1,000,000 and that accounts are filed with the Companies Registrar. Generally an individual shareholder's liability is limited to the amount of the share capital he is required to subscribe.

Advantages

In view of the problems and costs of incorporating an existing business, it is important to try and select the correct trading medium at the commencement of operations. It is not true to say that every business should start life as a company.

Many businesses are carried on in a safe and efficient manner by sole traders or partnerships. Whilst recognising the possible commercial advantages of a limited company, taxation advantages exist for sole traderships and partnerships, such as income tax deferral and National

Insurance saving. No decision should be taken without first seeking professional advice.

The benefit of limited liability should not be ignored although this can largely be negated by banks seeking personal guarantees. In addition, it may be easier for the companies to raise finance because the bank can take security on the debts of the company that could be sold in the future, particularly if third-party finance has been obtained in the form of equity.

Self-assessment

From the tax year 1996/97 the burden of assessing tax shifted from the Inland Revenue to the individual tax payer. The main features of this system are as follows:

- the onus is on the taxpayer to provide information and to complete returns
- tax will be payable on different dates
- the taxpayer has a choice: he can calculate his tax liability at the same time as making his return and this will need to be done by 31st January following the end of the tax year. Alternatively, he can send in his tax return before 30 September and the Inland Revenue will calculate the tax to be paid on the following 31 January
- the important aspect to the system is that if the return is late, or the tax is paid late, there will be automatic penalties and/or surcharges imposed on the taxpayer.

Tax correspondence

Businessmen do not like letters from the Inland Revenue but they should resist the temptation to tear them up or put them behind the clock and forget about them. All Tax Calculations and Statements of Account should be checked for accuracy immediately and any queries should be put to your accountant or sent to the Tax District that issued the document.

Keep copies of all correspondence with the Inland Revenue. Letters can be mislaid or fail to be delivered and it is essential to have both proof of what was sent as well as a permanent record of all correspondence.

Dates tax due

Income Tax

Payments on account (based on one half of last year's liability) are due on 31 January and 31 July. If these are insufficient there is a balancing payment due on the following 31 January – the same day as the tax return needs to be filed. For example:

for the year 2005/06	Tax due £5,000 (2004/05 was £4,000) First payment on account of £2,000 is due on 31.01.06 Second payment on account of £2,000 is due on 31.07.06 Balancing payment of £1,000 is due on 31.01.07

Note that on 31.01.07, the first payment on account of £2,500 falls due for the tax year 2006/07.

Tax in business

Spouses in business

If spouses work in the business, perhaps answering the phone, making appointments, writing business letters, making up bills and keeping the books, they should be properly remunerated for it. Being a payment to a family member, the Inspector of Taxes will be understandably cautious in allowing remuneration in full as a business expense. The payment should be:

- actually paid to them, preferably weekly or monthly and in addition to any housekeeping monies
- recorded in the business book
- reasonable in amount in line with their duties and the time spent on them.

If the wages paid to them exceed £94.00 per week, Class 1 employer's and employee's NIC becomes due and if they exceed £4,895 p.a. (assuming they have no other income) PAYE tax will also be payable.

It should also be noted that once small businesses are well established and the spouses' earnings are approaching the above limits, consideration may be given to bringing them in as a partner. This has a number of effects:

- there is a reduced need to relate the spouse's income (which is now a share of the profits) to the work they do
- they will pay Class 2 and Class 4 NIC instead of the more costly Class I contributions and PAYE will no longer apply to their earnings but remember that, as partners, they have unlimited liability.

Premises

Many small businessmen cannot afford to rent or buy commercial premises and run their enterprises from home using part of it as an office where the books and vouchers, clients' records and trade manuals are kept and where estimates and plans are drawn up. In these circumstances, a portion of the outgoings on the property may be claimed as a business expenses. An accountant's advice should be sought to ensure that the capital gains tax exemption that applies on the sale of the main residence is not lost.

Fixed Profit Car Scheme

It may be advantageous to calculate your car expenses using a fixed rate per business mile. A condition is that your annual turnover is below the VAT threshold (currently £60,000). Ask your accountant about this. A proper record of business mileage must be kept.

Vehicles

Car expenses for sole traders and partners are usually split on a fractional mileage basis between business journeys, which are allowable, and private ones, which are not, and a record of each should he kept. If the business does work only on one or two sites for only one main contractor, the inspector may argue that the true base of operations is the work site not the residence and seek to disallow the cost of travel between home and work. It is tax-wise and sound business practice to have as many customers as possible and not work for just one client.

Business entertainment

No tax relief is due for expenditure on business entertainment and neither is the VAT recoverable on gifts to customers, whether they are from this country or overseas. However, the cost of small trade gifts not exceeding £10 per person per annum in value is still admissible provided that the gift advertises the business and does not consist of food, drink or tobacco.

Income tax (2005/06)

Personal allowances

The current personal allowance for a single person is £4,895. The personal allowance for people aged 65 to 74 and over 75 years are £7,090 and £7,220 respectively. The married couple's allowance was withdrawn on 5 April 2000, except for those over 65 on that date.

Taxation of husband and wife

A married woman is treated in much the same way as a single person with her own personal allowance and basic rate band. Husband and wife each make a separate return of their own income and the Inland Revenue deals with each one in complete privacy; letters about the husband's affairs will be addressed only to him and about the wife's only to her unless the parties indicate differently.

Rates of tax

Tax is deducted at source from most banks and building societies accounts at the rate of 20%. The rates of tax for 2005/06 are as follows:

Lower rate: 10% on taxable income up to £2,090
Basic rate: 22% on taxable income between £2,090 and £32,400
Higher rate: 40% on taxable income over £32,400

Dividends carry a 10% non-repayable tax credit. Higher rate taxpayers pay a further tax on dividends of 22.5%.

Mortgage interest relief

This is no longer available after 5 April 2000.

Business losses

These are allowed only against the income of the person who incurs the loss. For example, a loss in the husband's business cannot be set against the wife's income from employment.

Joint income

In the case of joint ownership by a husband and wife of assets that yield income, such as bank and building society accounts, shares and rented property, the Inland Revenue will treat the income as arising equally to both and each will pay tax on one half of the income. If, however, the asset is owned in unequal shares or one spouse only and the taxpayer can prove this, then the shares of income to be taxed can be adjusted accordingly if a joint declaration is made to the tax office setting out the facts.

Capital Gains Tax

Where an asset is disposed of, the first £8,500 of the gain is exempt from tax. In the case of husbands and wives, each has a £8,500 exemption so if the ownership of the assets is divided between them, it is possible to claim exemption on gains up to £17,000 jointly in the tax year. Any remaining gain is chargeable as though it were the top slice of the individual's income; therefore according to his or her circumstances it might be charged at 10%, 20% or 40%.

Self-employed NIC rates (from 6 April 2005)

Class 2 rate
Charged at £2.10 per week. If earnings are below £4,895 per annum averaged over the year, ask the DSS about 'small income exception'. Details are in leaflet CA02.

Class 4 rate
Business profits up to £4,895 per annum are charged at NIL. Annual profits between £4,895 and £32,760 are charged at 8% of the profit. There is a charge on profits over £32,760. Class 4 contributions are collected by the Inland Revenue along with the income tax due.

Capital allowances (depreciation) rates

Plant and machinery:	25% (50% first-year allowance is available for certain small businesses)
Business motor cars - cost up to £12,000:	25%
- cost over £12,000:	£3,000 (maximum)
Industrial build	4%
Commercial and industrial buildings in Enterprise Zones:	100%
Computers and software equipment	100%

THE CONSTRUCTION INDUSTRY TAX DEDUCTION SCHEME

General

The new Construction Industry Tax Deduction Scheme is known as the 'CIS' scheme and replaced the old '714' scheme. As the scheme operates whenever a contractor makes a payment to a sub-contractor, the businessman should visit his local income tax enquiry office and obtain copies of the Inland Revenue booklet IR 14/15 (CIS) and leaflet IR 40 which will explain the conditions under which the Inland Revenue will issue a registration card or (CIS6) certificate and precisely when the scheme applies.

Everyone who carries out work in the Construction Industry Scheme must hold a registration card (CIS4) or a tax certificate (CIS6). Certain larger companies use a special certificate (CIS5).

If the sub-contractor has a registration card but does not hold a valid tax certificate (CIS6) issued to him by the Inland Revenue, then the contractor *must* deduct 18% tax from the whole of any payment made to him (excluding the cost of any materials) and to account to the Inland Revenue for all amounts so withheld.

To enable the subcontractor to prove to the Inspector of Taxes that he has suffered this tax deduction, the contractor must complete the three-part tax payment voucher (CIS25) showing the amount withheld. These vouchers must be carefully filed for production to the Inspector after the end of the tax year along with the tax return. Any tax deducted in this way over and above the sub-contractor's agreed liability for the year will be repaid by the Inland Revenue. If he holds a (CIS6) certificate the payment may be made in full without deducting tax.

A small business that does work only for the general public and small commercial concerns is outside the scheme and does not need a certificate to trade. If, however, it engages other contractors to do jobs for it, the business would have to register under the scheme as a contractor and deduct tax from any payment made to a sub-contractor who did not produce a valid (CIS6) certificate. If in doubt, consult your accountant or the Inland Revenue direct.

VAT

The general rule about liability to register for VAT is given in the VAT office notes. It is possible to give here only a brief outline of how the tax works. The rules that apply to the construction industry are extremely complex and all traders must study *The VAT Guide* and other publications.

Registration for VAT is required if, at the end of any month, the value of taxable supplies in the last 12 months exceeds the annual threshold or if there are reasonable grounds for believing the value of the taxable supplies in the next 30 days will exceed the annual threshold.

Taxable supplies include any zero-rated items. The annual threshold is £60,000. The amount of tax to be paid is the difference between the VAT charged out to customers *(output tax)* and that suffered on payments made to suppliers for goods and services *(input tax)* incurred in making taxable supplies. Unlike income tax there is no distinction in VAT for capital items so that the tax charged on the purchase of, for example, machinery, trucks and office furniture, will normally be reclaimable as *input tax*.

VAT is payable in respect of three monthly periods known as 'tax periods'. You can apply to have the group of tax periods that fits in best with your financial year. The tax must be paid within one month of the end of each tax period. Traders who receive regular repayments of VAT can apply to havthem monthly rather than quarterly. Not all types of goods and services are taxed at 17.5% (i.e. the standard rate). Some are exempt and others are zero-rated.

Zero-rated

This means that no VAT is chargeable on the goods or services, but a registered trader can reclaim any *input* tax suffered on his purchases. For instance, a builder pays VAT on the materials he buys to provide supplies of constructing but if he is constructing a new dwelling house, this is zero rated. The builder may reclaim this VAT or set it off against any VAT due on standard rated work.

Exempt

Supplies that are exempt are less favourably treated than those that are zero rated. Again no VAT is chargeable on the goods or services but the trader cannot reclaim any *input* tax suffered on his purchases.

Standard-rated

All work which is not specifically stated to be zero rated or exempt is standard-rated, i.e. VAT is chargeable at the current rate of 17.5% and the trader may deduct any *input* tax suffered when he is making his return to the Customs and Excise. If for any reason a trader makes a supply and fails to charge VAT when he should have done so (e.g. mistakenly assuming the supply to be zero rated), he will have to account for the VAT himself out of the proceeds. If there is any doubt about the VAT position, it is safer to assume the supply is standard rated, charge the appropriate amount of VAT on the invoice and argue about it later.

Time of supply

The *time* at which a supply of goods or services is treated as taking place is important and is called the 'tax point'. VAT must be accounted for to the Customs and Excise at the end of the accounting period in which this 'tax point' occurs. For the supply of goods which are 'built on site', the 'basic tax point' is the date the goods are made available for the customer's use, whilst for *services* it is normally the date when all work except invoicing is completed.

However, if you issue a tax invoice or receive a payment before this 'basic tax point' then that date becomes a tax point. In the case of contracts providing for stage and retention payments, the tax point is either the date the tax invoice is issued or when payment is received, whichever is the earlier.

All the requirements apply to sub-contractors and main contractors and it should be noted that, when a contractor deducts income tax from a payment to a sub-contractor (because he has no valid CIS6) VAT is payable on the full gross amount *before* taking off the income tax.

Annual accounting

It is possible to account for VAT other than on a specified three month period. Annual accounting provides for nine equal installments to be paid by direct debit with annual return provided with the tenth payment. £300,000.

Cash accounting

If turnover is below a specified limit, currently £600,000, a taxpayer may account for VAT on the basis of cash paid and received. The main advantages are automatic bad debt relief and a deferral of VAT payment where extended credit is given.

Bad debts

Relief is available for debts over 6 months.

Part Four

GENERAL DATA

GENERAL CONSTRUCTION DATA

The metric system

Linear

1 centimetre (cm)	=	10 millimetres (mm)
1 decimetre (dm)	=	10 centimetres (cm)
1 metre (m)	=	10 decimetres (dm)
1 kilometre (km)	=	1000 metres (m)

Area

100 sq millimetres	=	1 sq centimetre
100 sq centimetres	=	1 sq decimetre
100 sq decimetres	=	1 sq metre
1000 sq metres	=	1 hectare

Capacity

1 millilitre (ml)	=	1 cubic centimetre (cm3)
1 centilitre (cl)	=	10 millilitres (ml)
1 decilitre (dl)	=	10 centilitres (cl)
1 litre (l)	=	10 decilitres (dl)

Weight

1 centigram (cg)	=	10 milligrams (mg)
1 decigram (dg)	=	10 centigrams (mcg)
1 gram (g)	=	10 decigrams (dg)
1 decagram (dag)	=	10 grams (g)
1 hectogram (hg)	=	10 decagrams (dag)

Conversion equivalents (imperial/metric)

Length

1 inch	=	25.4 mm
1 foot	=	304.8 mm
1 yard	=	914.4 mm
1 yard	=	0.9144 m
1 mile	=	1609.34 m

Area

1 sq inch	=	645.16 sq mm
1 sq ft	=	0.092903 sq m
1 sq yard	=	0.8361 sq m
1 acre	=	4840 sq yards
1 acre	=	2.471 hectares

Liquid

1 lb water	=	0.454 litres
1 pint	=	0.568 litres
1 gallon	=	4.546 litres

Horse-power

1 hp	=	746 watts
1 hp	=	0.746 kW
1 hp	=	33,000 ft.lb/min

Weight

1 lb	=	0.4536 kg
1 cwt	=	50.8 kg
1 ton	=	1016.1 kg

Conversion equivalents (metric/imperial)

Length

1 mm	=	0.03937 inches
1 centimetre	=	0.3937 inches
1 metre	=	1.094 yards
1 metre	=	3.282 ft
1 kilometre	=	0.621373 miles

Area

1 sq millimetre	=	0.00155 sq in
1 sq metre	=	10.764 sq ft
1 sq metre	=	1.196 sq yards
1 acre	=	4046.86 sq m
1 hectare	=	0.404686 acres

Liquid

1 litre	=	2.202 lbs
1 litre	=	1.76 pints
1 litre	=	0.22 gallons

Horse-power

1 watt	=	0.00134 hp
1 kw	=	134 hp
1 hp	=	0759 kg m/s

Weight

1 kg	=	2.205 lbs
1 kg	=	0.01968 cwt
1 kg	=	0.000984 ton

Temperature equivalents

In order to convert Fahrenheit to Celsius deduct 32 and multiply by 5/9.
To convert Celsius to Fahrenheit multiply by 9/5 and add 32.

Fahrenheit	**Celsius**
230	110.0
220	104.4
210	98.9
200	93.3
190	87.8
180	82.2
170	76.7
160	71.1
150	65.6
140	60.0
130	54.4
120	48.9
110	43.3
100	37.8
90	32.2
80	26.7
70	21.1
60	15.6
50	10.0
40	4.4
30	-1.1
20	-6.7
10	-12.2
0	-17.8

Areas and volumes

Figure	**Area**	**Perimeter**
Rectangle	Length × breadth	Sum of sides
Triangle	Base × half of perpendicular height	Sum of sides
Quadrilateral	Sum of areas of contained triangles	Sum of sides
Trapezoidal	Sum of areas of contained triangles	Sum of sides
Trapezium	Half of sum of parallel sides × perpendicular height	Sum of sides
Parallelogram	Base × perpendicular height	Sum of sides
Regular polygon	Half sum of sides × half internal diameter	Sum of sides
Circle	pi × radius^2	pi × diameter or pi × 2 × radius

Figure	**Surface area**	**Volume**
Cylinder	pi × 2 × radius^2 × length (curved surface only)	pi × radius^2 × length
Sphere	pi × diameter^2	Diameter3 × 0.5236

Mild steel cisterns

Length mm	Width mm	Depth mm	Capacity mm
457	305	305	18
610	305	371	36
610	406	371	54
610	432	432	68
610	457	482	86
686	508	508	114
736	559	559	159
762	584	610	191
914	610	584	227
914	660	610	264

Plastic cisterns

Ref.	Capacity litres	Capacity gallons	Weight kg
PC4	18	4	0.85
PC15	68	15	2.95
PC25	114	25	3.40
PC40	182	40	6.35

Roof drainage

	Area m2	Pipe mm	Gutter mm
One end outlet	15	50	75
	38	68	100
	100	110	150
Centre outlet	30	50	75
	75	68	100
	200	110	150

Index

Taylor & Francis Group
A division of T&F informa plc